+65p14

*The Development and
Organization of Scientific Knowledge*

The Development and Organization of Scientific Knowledge

Harold Himsworth
*formerly Professor of Medicine
in the University of London
and Secretary of the Medical Research Council*

HEINEMANN: LONDON

William Heinemann Ltd
LONDON MELBOURNE TORONTO
JOHANNESBURG AUCKLAND

First published 1970
© Harold Himsworth 1970
434 33630 0

Printed in Great Britain by
Western Printing Services Ltd, Bristol

CONTENTS

		Page
Foreword		vii
CHAPTER 1	Scientific Development and its Social Significance	1

THE STRUCTURE OF SCIENTIFIC KNOWLEDGE

CHAPTER 2	Biomedicine	7
CHAPTER 3	Sequences and Kinds of Knowledge	13
CHAPTER 4	Agriculture, Materials and Energy	26
CHAPTER 5	The Structure of Scientific Knowledge	46

THE ORGANIZATION OF SCIENTIFIC DEVELOPMENT

CHAPTER 6	The Evolution of Organized Research	65
CHAPTER 7	The Universities and Research	75
CHAPTER 8	Central Research Organizations	98
CHAPTER 9	Research in the Professions and Industry	115
CHAPTER 10	Research Policy and Provinces of Knowledge	124
CHAPTER 11	Organization and Research Policy	142

CONTENTS

CHAPTER 12 *National Policy and Scientific Development*   153

CHAPTER 13 *Epitome*   171

Index   177

FOREWORD

IN WRITING THIS ESSAY I have been very conscious of my limitations. My scientific experience is limited to one province of knowledge, the biomedical, and my administrative experience largely confined to one country, the United Kingdom. I am, therefore, very well aware of the inadequacies that must inevitably occur in this attempt to deal with such a vast problem as the structure and organization of scientific knowledge. To some extent I have deliberately accepted limitations. I have not for instance taken organizational considerations beyond the national level. As a result I have not discussed either the great achievements or increasing significance of the developing international bodies such as the World Health Organization. This is not because I yield to anybody in my appreciation of these but because I am primarily concerned to show the relationship between concepts and action; and for this purpose the universally familiar unit of the national scale was the most appropriate in the period in which we are now living. Being essentially a physician, I am also very much aware of my ignorance of other fields of scientific knowledge. Nevertheless such ignorance is the common lot of scientists in any field today and, perhaps, biomedical science has as wide relations with scientific knowledge in general as any other. Be that as it may, having had unusual opportunities to see my particular field as a whole and having been brought in contact with policy at the national level which concerned all fields of scientific development, I eventually decided to put my thoughts on paper.

It would be impossible for me to express my indebtedness, even if I could now identify this, to all the people with whom I have

FOREWORD

discussed these problems over many years. I must, however, record how much I owe to Sir Landsborough Thomson who, with his unrivalled experience of the early developments of scientific organization in this country, first introduced me to the administrative aspects. In regard to this present essay, I would particularly wish to acknowledge my debt to Sir Charles Harington, Sir Aubrey Lewis and Sir Frank Turnbull who have read the earlier draft and have given me the benefit of their informed criticism. Many of their suggestions I have adopted in part or in whole. But I must take entire responsibility for the views put forward and for the way that they are presented.

Finally, I would like to thank the Controller of Her Majesty's Stationery Office for permission to quote extensively from various official publications.

London. 24th April 1969

* I *

SCIENTIFIC DEVELOPMENT AND ITS SOCIAL SIGNIFICANCE

DURING THE LAST QUARTER OF A CENTURY the position of science in public estimation has changed significantly. From its status in general regard as a branch of learning that could also yield results of great practical importance, it has received general recognition as an essential and utilizable means for the purposive development of our civilization. It has thereby become a major concern of social policy and questions relating to its future development have become matters not only of academic but also of public interest.

In theory, these changes have long been foreseen as inevitable. Over the last century and a half, more and more scientists have been insisting on the increasing significance of their developing knowledge for national policies, and far-sighted administrators like Haldane, Addison and Morant have appreciated the changes in our traditional machinery of government that the integration of such expert knowledge would entail. As a result of experiences in the First World War, the bearing of scientific knowledge on national purposes came increasingly to be recognized by governments. It required, however, the experience of the Second World War, and particularly the development of the atomic bomb, for the public in general to realize fully the extent to which scientific development had now become vital to their interests and that the status of their particular community, or even their own survival, was bound up with its continued progress. Thereafter, in all developed countries, support for scientific research escalated, scientific progress became a major national concern, and expectation rather than hope became the public measure of scientific achievement.

It is difficult, however, to avoid the impression that the

scientific community in general was not wholly prepared for the speed and scale with which these changes occurred and still less for their implications. Naturally, the increased support for scientific research and the increasing prestige of scientific opinion were welcomed unreservedly, even if their adequacy did not go unquestioned. But distrust of authority is inherent in all creative workers. They feel that their work is largely insusceptible to organization and that it may be destroyed by attempts to confine it within a conventional framework. Yet it was inescapable that increased public support for science implied increased justification to public authority. Further, that the plans and arrangements for its developments would now need to convince an audience whose ideas on organization and purpose derived from other and different spheres of activity. The scientific community was in consequence faced with a dilemma. On the one hand were the continually rising demands of research for support, with their inescapable need for justification; on the other the fear that, by accepting such support, research might be consigning itself to conditions in which its future development would be impeded or even impossible. The widespread attempts by scientists to formulate what is called Scientific Policy and the attention that they are now giving to the organization of research, is the expression of their concern with this problem.

The problems posed for the administrator have been different but similarly challenging to previous assumptions. In modern society, dependent as it is on expert knowledge, it is often bordering on an euphemism to call scientific opinion 'advice'. Frequently it is for all practical purposes policy itself. For a tradition of organization that has assumed experts to be no more than advisers, the changed position of science raises genuine constitutional issues. But the administrative planner is not entirely without precedent. In respect of the military services, society has always been faced with the problem of reconciling dependence upon individual expert judgment with the requisite delegation of executive authority and it may well be that, by considering their analogy, some useful pointers to the arrangements now required to relate science to government may be discovered.

But it would be idle to pretend that, despite this activity in both scientific and administrative circles, any country has as yet solved the problems posed by the emergence of science as a recognized social force. It is clear, however, that planning must take its start from a consideration of the nature of scientific knowledge and the requirements of the creative activity needed to advance it. Only on this basis can we see what concepts in our traditional arrangements are essential and what are no more than legacies from past circumstances. If there were an agreement on a structure of scientific knowledge, showing the relation of the different subjects and disciplines to each other, it would then be possible to see the form of the organization required to give it expression. If, further, it were possible to show a natural sequence in the intellectual relationships of a series of subjects, we should have a logical basis for policy. Without clear ideas on these points, however, our planning is condemned to expediency. Unfortunately, owing to the high degree of specialization in research, every subject tends to be regarded by those who engage in it as the centre or fulcrum of the scientific universe, or, at least, as of more importance than is generally conceded. In such circumstances agreement is difficult. That is not to say, however, that such relationships and such sequences do not exist nor that, if they do, they could not form the basis of a rational policy for science both as regards its internal organization and also as regards its social relationships. It is my belief that they do indeed exist and that, although it is difficult to distinguish them if one begins by looking at science as one whole, they can readily be seen if the approach is made through particular fields of scientific activity.

Our ideas on the structure of scientific knowledge were formulated in a far simpler era and under the influence of philosophies that have now lost much of their authority. Today, the range and complexity of such knowledge have increased beyond anything that could then have been conjectured. It may well be, therefore, that the conceptual approximations that previously sufficed are no longer sufficient to meet the more exacting conditions of the present. This I sincerely believe is the situation and, in my view, many of our present difficulties derive from this cause. It seems to me, therefore, that, if we are to provide for the future, the first step is for us, as scientists, to reappraise the ideas that we traditionally hold about the structure and development of our knowledge. To this end I propose first to examine the province of knowledge with which I am most familiar, biomedical research, and then, in the light of the indications from this, to look at other provinces to see if there is a structure common to them all. Thereafter, in the second part of this essay, I shall, on the basis of the conclusions reached in the first, consider the instruments at our disposal for the advancement of scientific knowledge and the question of policy for its further development.

The Structure of Scientific Knowledge

* 2 *

BIOMEDICINE

I

THE UNIVERSAL DESIRE OF MEN not to be ill, or to die before they must, has always ensured for medical knowledge a high place in public estimation. Until relatively recent times, however, understanding was so meagre that illness was accepted with resignation as the common lot and recovery from it as a matter for thankfulness. But over the last two centuries the situation has increasingly changed. On the basis of the long accumulation of empirical knowledge of disease and the fumbling evolution of concepts to provide a rational basis for this, understanding has developed and, with increasing frequency, particular diseases have been brought under control. At first slow, this progression has rapidly gathered momentum so that it is now no exaggeration to say that medical knowledge has advanced more in the last century (and more particularly in the last fifty years) than in all its previous history. It is only necessary to glance at the headlines of the practical results of this modern knowledge to get some measure of the revolution in outlook and attitude that has recently occurred.

Anaesthesia, antiseptics and later aseptic techniques have transformed surgery from a desperate recourse into a wide-ranging measure of amelioration. Immunization by vaccines or inoculation has given us the means to prevent smallpox, diphtheria, tetanus, yellow fever, poliomyelitis and substantially to mitigate the prevalence and severity of several other common infections. The chemo-prophylaxis of malaria has made living safe in the tropics. Knowledge of nutrition has eliminated the gross deficiency diseases in all countries that have the resources to apply it. In this country over the last fifty years, the death rate in the first year of life has fallen from more than 100 out of every 1,000 children born alive to under 20. But perhaps the most dramatic single achievement has been the control of infections by sulphonamides and antibiotics. To have seen lobar pneumonia cut short by sulphapyridine, and bacterial endocarditis resolve under penicillin, still excites wonder in physicians of my generation. There are many thousands of people alive and well today who, but for the introduction of effective drugs for the treatment of tuberculosis in the last two decades, would have been dead. What penicillin and the later antibiotics have meant in terms of saving life, preventing illness and increasing national productivity, is beyond conjecture.

And alongside these dramatic successes is the progress in alleviation which, from the point of view of the individual patient, is still the most important role of medicine. By definition, alleviation applies to conditions that we can neither cure nor prevent. It is, therefore, as wide-ranging as medicine itself. The sophistication in knowledge now required can be seen if one considers the control of high blood pressure, the modern management of heart or kidney failure, the chemotherapy and management of mental illness, radiotherapy for cancer, the control of conditions like diabetes, and the transplantation of organs.

Much, if not most, of the progress of which the above examples are an indication has occurred in the lifetime of men now living. But even if measured in terms of achievement, it is obviously only a beginning. There are wide ranges of conditions in which the record is still largely one of failure and some of the advances that have been made could well be threatened by changing circumstances. Without question, however, progress has already been so substantial and conspicuous that it has altered, both in the estimation of the public and of research workers in this field, the whole climate of expectation with regard to the promise of biomedical research. If this expectation is to be satisfied it will be necessary, however, to have a clear idea of the range and depth of knowledge that such research must now encompass.

BIOMEDICINE

II

All disease is a deviation from the normal. It is for this reason that research on disease cannot be divorced from research on the normal, and why the whole is more properly described as 'biomedical' rather than 'medical'. Strategy in biomedical research consists, therefore, primarily in trying to keep the medical and biological wings moving forward in alignment together. What this means in practice can best be seen from actual examples, and as the first example we can take cancer research.

Historically, cancer research began with observations on sick persons. Thereby the types of the disease and their natural history, and the descriptive pathology of cancer were defined. In such work elusive glimpses of possible associations with environmental factors began to emerge and, from these clues, epidemiological studies, aimed at validating or rejecting such indications, arose. Experimental pathology now came into the picture. Rational attempts became possible to produce tumours in normal animals by exposing them to agents, such as chemicals, radiations or viruses, that had been identified by clinical and epidemiological studies. The ability to produce tumours experimentally immediately widened the approach. The range of substances, physical forces and living agents capable of initiating cancer could now be actively investigated and the features they had in common sought. The susceptibility of different strains of animals to such agents and the factors bearing on this could be explored. The question what processes are interfered with in normal tissues so as to make them cancerous could now begin to be asked. Inevitably biochemistry, chemistry, radiobiology, virology and genetics were drawn in. Concern with the cell, and particularly the mechanisms that maintain its functional normality, became inescapable. Radiations damage chromosomes and produce mutations, as do several carcinogenic chemicals; viruses are packets of nucleic acids that can transform the whole function of the cell. Research on microbial genetics, the structure and function of nucleic acids, the control of intracellular processes—in short, biology at the molecular level—has now become essential.

Cancer research is no exceptional example. Similar logical sequences of necessary knowledge occur throughout the biomedical field.

The first step in the understanding of microbial infections is to define the clinical illness and its descriptive pathology. Until this is done, epidemiology cannot begin and attempts to isolate a causative agent are unlikely to lead to anything but confusion. Once this has been accomplished, however, and a uniform population of cases can be studied, the possibility of identifying a microbial factor common to all can be entertained. Thereafter, the production of the illness experimentally in animals becomes feasible and more controllable investigation is thereby possible. Research on the reaction of the body to infection, determination of the metabolism of the micro-organisms and their response to foreign substances, such as drugs, follows. A sequence of effort in clinical medicine, pathology, microbiology, immunology, biochemistry and pharmacology comes into play. And it may not stop there. Changes in the metabolism of the micro-organisms and their susceptibility to chemotherapeutic agents, alterations in their virulence, may well lead to work at the molecular level on their intracellular processes.

As a further example, consider the endocrine glands, their disorders and functions.

The endocrine glands are organs that secrete into the blood stream. The chief ones are the sex glands, the thyroid, the adrenals, the pituitary and the insulin-producing islets of the pancreas; and the normal development and functioning of the body depends vitally upon their secretions (hormones) being maintained at normal levels. Knowledge in this field not only took its start, but is largely based, upon investigation of the effects of over-supply and under-supply of the substances that they secrete, and the motive influence in its development was the study of human disease due to the natural experiments produced by over- or under-functioning of particular glands. Giants have excited interest from time immemorial. A classical example of one kind is given in the biblical story of Samson with its description of phenomenal growth and strength and the subsequent development of blindness and impotence. The first step was the recognition of this type of gigantism and its association with a non-cancerous tumour of the pituitary gland. Later, extracts of such tumours, or, even more significantly, of the normal gland, were shown to produce a great increase in the growth of young normal animals. The 'growth hormone' was thereby discovered and it was speedily evident that this had profound effects upon biochemical processes within living cells. A flood of investigations, now stretching down to the molecular level, was thus set in train, and it is clearly evident that these will lead to a great increase in our understanding not only of the disease that initiated thought in this field, but also of the control of activity within the normal cell itself.

As a last example, let us look at the development of our knowledge of nutrition.

Man has always recognized his dependence on an adequate supply of food, but, although he probably suspected the relevance of quality, it was only in relatively recent times, with the understanding of the diseases due to vitamin deficiency, that he came to appreciate its importance. Scurvy, rickets, pellagra and beri-beri have been with us always and the development of our knowledge on the latter will serve as an illustration.

The first step was the differentiation of beri-beri from other forms of heart failure and neurological disease and its further definition by pathology. At that stage, progress halted until it was noted that the disease was prevalent if the main constituent of the diet was polished rice but not if the rice was unpolished. The way was thus open to the experimental production of the condition in animals, when it was speedily shown that this could be prevented or cured by an extract of rice polishings. Biochemistry and chemistry could now come into operation. Over a period of years the anti-beri-beri substance was isolated (and the name 'vitamin' coined for this class of substances), identified and synthesized. The point of its participation in the biochemical economy of the body was determined and, in the course of this, a flood of light was thrown on the vital intracellular molecular interactions. All this, from its start to the present advanced stage, took place within a framework of reference stemming from nature's experiment in producing the disease beri-beri, by which the validity of each step in the progressively deeper studies could be checked.

Examples of such sequences could be multiplied many times in the province of biomedical knowledge. That each represents an assembly of disciplines necessary for the solution of the initial problem and that there is something like an order of succession in the way that the different disciplines can profitably be brought to bear, is evident. But is this all the significance that attaches to these sequences. Are they simply a deployment of different knowledges in a particular way for some particular *ad hoc* utilitarian purpose, or have they a deeper meaning? Are they merely expedients, or are they the valid reflection in thought of relationships that exist in the ordering of natural phenomena themselves?

This is an important question. According as to how we answer it, we shall arrive at very different ideas about the structure of scientific knowledge and the course of its development. Clearly, however, the first consideration is whether sequences of this kind are a peculiarity of biomedical knowledge or whether they are a general feature in all provinces of scientific knowledge. If examination shows that the former is the case, then their significance is limited. If, however, it shows that they are a general feature of all provinces of knowledge, then we can reasonably conclude that they are of key importance in any consideration of the structure of scientific knowledge and all that follows from this. But before addressing ourselves to such an examination, it is necessary first to look in a little more detail at the characteristics of such sequences.

3
SEQUENCES AND KINDS OF KNOWLEDGE

I

LOOKING AT THE KINDS OF SEQUENCES that we have been considering, there are three features that are apparent. The first is that within each there is, intellectually speaking, no break. From the extreme of a specialized interest in the clinical field on the one hand to unspecialized studies, even at the molecular level, on the other, each sequence forms one smooth intellectual continuum.[1] At no place can we say that knowledge on the one side is divorced from that on the other or that there is some obstacle preventing the flow of knowledge in either direction.

The second feature is practically a corollary of the first. It is that within such intellectual continua each subject is in context. Each is so placed as to have subjects on either side of it with purposes contiguous to its own and which contribute information that is relevant to its own purposes.

The third feature is that the order in which the different subjects occur along such biomedical sequences (and therefore within the corresponding intellectual continuum) is broadly the same: clinical, pathological, physiological, biochemical, and so on down to the molecular level.

These considerations apply individually to all such sequences, but, looking at the sequences collectively, a further important

---

[1] The term 'specialized' (in relation to knowledge) will be used in this essay to mean knowledge of specialized situations; the term 'unspecialized' to mean knowledge that is a common element in a number of specialized situations. Thus, knowledge of a disease state due to deficiency of a particular enzyme would be specialized. Knowledge of the action of that enzyme and its role in normal function, less specialized. Knowledge of the general properties of enzymes substantially unspecialized. (See also pp. 59 to 61.)

feature emerges. As each proceeds from its specialized towards its unspecialized extreme, each comes to have progressively more ground in common, each to include subjects that are increasingly more capable of generalization beyond the particular sequence.

At the specialized extreme of different sequences, knowledge is so orientated to a particular purpose that it is of relatively little moment to others. For example, there is little in common between the clinical features or pathology of say a cancer of the lung and those of an infectious illness like typhoid fever. It is hardly possible, however, to imagine any sequence in the biomedical province that does not, or will not sooner or later, include biochemistry. At the level where biochemistry operates, the partitions between biomedical sequences have become, so to speak, highly permeable, and knowledge and experience in this subject flow freely across between adjoining sequences. Nor is this phenomenon limited to one province. Adjoining the province of medical knowledge are those of agriculture and those aspects of manufacture that are biologically orientated. Sequences in these will also include biochemistry and contribute likewise to the common knowledge of the subject. Similar considerations apply to other subjects in proportion to their distance from the specialized periphery. The spread of pathology is relatively limited, physiology somewhat greater, while the spread of physics and chemistry, with their wide range across many provinces of scientific knowledge, is wide in the extreme. As, therefore, one penetrates deeper from the specialized periphery through progressively less specialized knowledge, the successive subjects become more capable of generalization and more widely relevant. It was, in consequence, a very pertinent observation of a British Treasury Committee when, some forty years ago, it remarked that the three existing central research organizations of medicine, agriculture and industry all participated in developing the same 'primary sciences' but that each stretched out to deal with a distinctive world of experience.[1]

From the preceding considerations it appears possible to

[1] Committee of Civil Research: *Report of Research Coordination Subcommittee*, 1928, paras 22, 25, 27. H.M.S.O. 63-45, London.

extract a generalization, at least in respect of the province of biomedical knowledge. It is that, within this province as a whole, knowledge is structured from the highly specialized (or mission-orientated) through the progressively less specialized until it penetrates to levels which are largely unspecialized; the whole forming one smooth intellectual continuum from the specialized periphery to the approaches to the unspecialized centre. This is clearly not only a pragmatical model but a historical one in that it indicates the way in which knowledge in the particular province has been developed. Whether or not it has further significance must, however, be left until we have examined thinking with regard to the kinds of scientific knowledge.

II

Traditionally, it is customary to classify the kinds of scientific knowledge, or research, into basic (or pure or fundamental), applied and developmental. This classification, of course, also implies sequence. It is, however, a sequence of an essentially different nature from that which we have been considering. It is a sequence of attributes rather than kinds of experience, and its implications are considerable.

To say that some one thing is basic or fundamental to some second is to imply that this second thing is essentially determined by the first. Similarly, to describe one thing as the applied form of another is to suggest that its essential features can be derived from that other. But are these adequate descriptions of the subjects of science?

Take biochemistry for instance. At one time it was seriously contended that biochemistry was no more than a branch of organic chemistry, and that as such it had no claim to be separately distinguished. Surely, however, this was to mistake the whole issue. The object of biochemistry is to analyse biological phenomena in terms of chemistry; and the operative word in this context is *analyse*. From this point of view, an organic chemist who works on the structure of compounds that happen to be of biological interest is no more a biochemist than a metallurgist, who works on the structure of metals, is an engineer. Chemistry certainly provides the biochemist with his tools (and he must know how to use these and adapt them to his purpose), but it is biology which provides his problems. It is these problems, not the tools that are used, that generate the ideas and determine the intellectual content that are characteristic of the subject in question and are the basis of its title to separate identity.

It is evident that biochemistry, unlike subjects such as anatomy, could never have come into existence *ab initio*. Until physiology, the study of normal biological function, had progressed to the stage at which it could define and identify the problems that were susceptible to analysis in chemical terms, it could make no contact with chemistry. When, however, it developed to this stage and its first tentative gropings began to emerge in what was then called 'physiological chemistry', this became feasible. Chemistry, although it had necessarily been concerned with the composition of substances produced by living organisms, had arisen independently and for quite different purposes. It was already substantially developed. When, therefore, it became possible for biology to effect a junction with chemistry the new tools it required were, in principle, ready to hand.

The emergence of biochemistry might well serve as a prototype of one of the major ways in which scientific knowledge evolves. This particular process of evolution is, however, so important to scientific development that it merits further elaboration. We might, therefore, consider three further examples—two nearer the specialized periphery of the province of biomedical knowledge and one nearer the unspecialized centre—before venturing any further conclusions.

Illnesses in which disorders of body chemistry occur constitute an important part of clinical medicine. Of these, the so-called metabolic diseases are typical. In these conditions, a train of chemical events that underlies some essential function of the living organism is qualitatively or quantitatively abnormal. Knowledge in this field begins when such conditions are distinguished from others and identified as a metabolic disorder. It develops by analysis of the problem thereby defined in terms of biochemistry.

In this case, however, it is the clinical problem that poses the question and biochemistry that provides the tools. In a similar way a clinical problem may be identified as a disorder of function in some organ of the body, such as the heart or kidney. In that case, however, the clinical research worker will, at the present stage of development in biomedicine, draw rather on physiology for the choice of his tools.

At the other end of the scale we can consider problems in the field that is now usually called molecular biology. The use of crystallographic techniques and knowledge which evolved in the subject of physics has, in recent years, thrown great light on the structure of certain important organic molecules. At first sight these achievements, despite their magnificence, might not seem to involve any departure from the purely physical sphere of thought. But, even in the early stages, the idea of function was inherent in the work; the biological function that the molecule in question was known to subserve. Evidently concepts of biological function could never have arisen from pure physical knowledge. Later this became even more evident when the structure of D.N.A. had been clarified. The essential role of that substance in the genetic process had come down the sequence of biological knowledge from Avery's discovery that a kind of D.N.A. would produce an inheritable transformation of a relatively avirulent form of a particular micro-organism (the pneumococcus) into one of high virulence. Its involvement with the genetic mechanism of the cell was thus established. Further, microbiologists had steadily accumulated great amounts of information on the differing metabolic requirements of different genetic strains of microbes. On the basis of this information, the genetic code was worked out and the full significance of the structure of D.N.A. indicated. Again, the problems that gave significance to the work and which differentiated these new studies from a mere application of physical knowledge, came not from physics, but from biology.

These examples could be multiplied many times. Indeed, the theme they illustrate permeates the whole province of biomedical knowledge and is of general significance for any theory of its development. Essentially it points to three conclusions. The first is that a productive marriage cannot be effected between two subjects until both have passed a certain stage of maturity. The second is that the characteristics of the offspring of any such marriage are determined not by the more but by the less 'basic' subject. The third is that a new subject extends the area of natural experience in which the more 'basic' can increase its development.

To some it may well seem a detraction to assign to subjects that are traditionally described as 'basic' or 'fundamental' or 'pure' the role of tools; and certainly any such descriptions of their nature would, in itself, be inadequate to account either for their significance or for their development. To clarify this point, let us first consider the most 'pure' of all subjects, mathematics.

The concepts and techniques of mathematics are widely used throughout the biomedical subjects. Yet biomedical scientists would be the first to disclaim any title to being mathematicians and to admit that their activities in this respect were limited to appropriating existing mathematical knowledge for their own purposes. They would readily concede that such knowledge had developed independently of the biological or medical, although they might trace its development back through another province of knowledge to the need for mensuration. They would recognize, however, that when sufficient numerical data and problems had accumulated, the relationships between numbers became problems in themselves and that attempts to solve these made contributions to mathematical knowledge that were independent of its origins. Nevertheless in relation to the rest of science the significance of mathematics is that of a tool.

Again, take chemistry. This subject is used by workers throughout the biomedical province and, at the periphery of this at least, few would claim that they could do much more than understand its language and manipulate its knowledge. As we have previously noted, chemistry is one of those subjects that underlie not only mere sequences but whole provinces of knowledge. Its roots of origin are multiple and although the need for the extraction, isolation and analysis of medicaments from natural sources was one of these, it was only one. Far more important

SEQUENCES AND KINDS OF KNOWLEDGE

were the roots coming from the need for colours, to extract metals, to treat fabrics, to develop pottery and glass. But the chemist as chemist is not concerned with the uses to which his results are put. From the beginning, his concern has been with the nature and handling of materials. It was from the accumulation of such experience and such lore that the categories of substances and their interactions slowly emerged. It was on this basis that the chemical theory evolved. Throughout, the purpose of men, when thinking as chemists, has been the better understanding, and consequent control, of materials. It is materials, their interactions and properties, that have essentially set the chemist his problems, and it is the body of knowledge pertaining to these that we call chemistry.

It is not difficult to understand how on the basis of such considerations the belief could grow up that there were fundamental levels of scientific enquiry where subjects generated their own inspiration and became, so to speak, self-orientated. That such a belief is in part true is indisputable. But it is not the whole truth. New problems of the interaction of materials, the discovery of new materials in nature, the unforeseen behaviour of known materials under new conditions, remain sources of potent stimulation to the understanding and theory of chemists. Thus, if we are to appreciate the larger significance of a subject we must take into account not only the thought that it generates from consideration of its own data, but also that stimulated by contiguous experience.

In a sense, all scientific subjects (with the possible exception of mathematics) are ambivalent. Each looks both outward and inward for its inspiration. The nearer the subject to the specialized or mission-orientated end of the sequence, the more dependent is it on subjects closer to the natural environment for the problems and data from which it elaborates its characteristic knowledge. On the other hand, the nearer a subject is to the unspecialized end of the sequence, the more self-orientated it is and the more it depends on the manipulation of its own data for its development. But, with the doubtful exception of mathematics, not even the most unspecialized subject can afford in the interests of its

THE STRUCTURE OF SCIENTIFIC KNOWLEDGE

continued development entirely to dispense with the intellectual stimulation it derives from prepared data coming down through the relevant sequences that connect it with the specialized periphery of knowledge, or to forego the opportunities to put its theory to exacting test in a wide range of specialized experience. In no meaningful sense, therefore, can such sequences of knowledge be described in terms of basic, applied and developmental. On the contrary to do so would be to give a completely misleading picture of their interrelations and individual significance. All the subjects are set in intellectual continua and it is from these that they derive their larger significance. None can be divorced from these contexts without adversely affecting not only its own progress but the development of all.

To develop the intellectual continua in which subjects, from the most specialized to the most unspecialized, find their context would, therefore, seem to be of the first order of importance. Today, however, it can no longer be taken for granted that this will occur spontaneously.

III

In any field of scientific knowledge, enquiry necessarily starts with particular problems. Whether it is motivated by need or curiosity is immaterial, provided this is strong enough to lead to action. Then, as enquiry develops it progresses to deeper and deeper levels. In the past, when knowledge was less, it was possible for a single very able man to span the whole of this process from the initiating problem to the depths of enquiry then attainable and, further, to work effectively at all these levels. Now, this is seldom so. Increasingly men have to confine their attention to the level at which they are working or those immediately contiguous to it. Indeed, nowadays, a man may have known no other levels and, in consequence, have little awareness of the over-all intellectual context of which his own interests are part. Specialization of function, with its corollary specialization of interest, is today the rule rather than the exception among working scientists, and this has brought its own problems.

These developments have occurred at an increasing pace over the last hundred years so that today they are a pervasive influence in our thinking. But it is important to be clear as to the nature of the process. Specialization is simply an organizational expedient forced upon us by the limitations of human capacity. It has no intellectual significance whatever. Yet men are inevitably influenced in the general concepts they form by the particular situation in which they find themselves. Given specialization, such bias is inescapable, and in accepting opinions from men whose interests centre in a particular subject, or at least a particular level of scientific enquiry, due allowance needs to be made for this.[1]

In approaching any question relating to the structure of scientific knowledge it is, therefore, important to be clear as to the level of consideration at which this is being discussed. There are several such possible levels. Each is entirely legitimate for its own purpose but, naturally, appearance and perspective differ according to the level of the particular viewpoint and, if this is not appreciated, serious confusion is inevitable.

At the level of the individual research worker, attention is necessarily concentrated on the type of problem in which he is an expert. His purpose is to produce as perfect a piece of work as is humanly possible. To this end, all other considerations are secondary.

At the level of the leader of a research team, the emphasis begins to shift. Although he must necessarily be concerned with the interests of the individuals in the team, he has to look at these not only in their own rights, but also in the context of their particular contribution to the combined operation that is his purpose.

At the next level, that concerned with a whole sequence of

[1] 'Formal reports delineating the achievement and promise of various fields all tend to be isomorphic. It makes little difference whether the field is astronomy, physics or computers: its achievements have been outstanding, its promise superb and its needs and tastes very expensive. Nor is this surprising. Each report is prepared by dedicated members of a particular scientific community whose passions and aspirations, as well as knowledge, centre on a single field.' Alvin S. Weinberg: *Minerva*, 1965, 4, 3.

knowledge, the emphasis shifts still farther in this direction. Effective progress now depends upon all the necessary knowledge advancing in alignment. If one lags (or even more so, if one is missing), then the whole advance may be held back. Consideration now needs to span the whole range of subjects that is necessary to its purpose and, within this, to see the component subjects in perspective and in relation to each other.

At a farther level still, that of a whole province of natural knowledge, the prospect broadens still more. Concern is now with whole assemblies of adjoining sequences, each of which, at one extreme, is specialized in intent and all of which, as they approach their other extreme, come increasingly to have interests in common. Interrelations at all levels, promotion of common and increasingly unspecialized interests and questions of proportionate intellectual contribution, come to bulk ever larger in consideration, and appreciation of context becomes a *sine qua non*.

There is thus a progressive shift of emphasis as consideration moves through the different levels until, when it reaches the last, the whole perspective may have changed so much as to be almost foreign to an observer at the first. It is important, therefore, that we should know the level of consideration at which any conclusion regarding the structure of knowledge has been reached. It is clearly unreasonable to expect a man in the valley to have exactly the same view of the topography of a district as one up the mountain-side. Both views are necessary but their purposes are different. If our purpose is to arrive at a concept of the over-all structure of scientific knowledge, then it is essential that we not only enter on the matter at the appropriate level of consideration but also that we do not stray from it. And the appropriate level for this purpose is obviously that with the widest perspective. It is only in the context of consideration at this level that the whole can be seen and individuals at other levels can be reasonably confident that they are within the field of vision and their particular interests are unlikely to escape attention.

IV

At this point it will be as well to consider the question of motivation in relation to research, for this also must change with the level of consideration.

To suggest that the motive which impels a man to do research is consideration of the public benefit that might possibly accrue from his work would be to be unrealistically sentimental. Certainly, it is an additional gratification for a research worker to know that, if his work is successful, it will help to satisfy some human need. Indeed, such a consideration may, in the first place, have induced him to pay attention to a particular category of problem. When, however, it comes to the actual doing of research by an actual individual such considerations, if they have ever existed, recede into the background. And it is quite proper that this should be so. At the level of the individual research project the indispensable requirement is a dispassionate and objective consideration of the natural phenomena under investigation and for this the only effective motive is curiosity, interest, fascination (call it what you will) in the phenomena for their own sake. To take a personal example. When I was engaged in research on the disease diabetes mellitus the hope that, as a result of my work, I might some day benefit my patients was a thought that, I have to confess, only came back into my mind at rare and relaxed moments. What really drove me on when actually working was the sheer interest of the problems posed by the particular disease. At the level of the individual worker the essential motive force prompting to research is the desire for understanding for its own sake.

But as one leaves the level of the individual, a different consideration assumes increasing importance. This is the significance of the actual or potential results of the research for the achievement of some particular intellectual or practical purpose. Thus at the level of the director of a multidisciplinary team, although the individual members of his staff are preoccupied by the interest of their particular aspect of the team's work, he himself must also be concerned with the significance of their

THE STRUCTURE OF SCIENTIFIC KNOWLEDGE

individual results for the over-all objective with which the team as a whole is engaged. At the next level, that of a man or body of men concerned with developing a whole sequence of knowledge, the question of the intellectual significance of the contribution by the component disciplines becomes of even more importance. And at the still further level, where the concern is with a whole province of knowledge, the question of the potential significance of projected contributions comes to occupy the whole forefront of consideration.

Thus at the two extremes of level, considerations are necessarily different. At the individual level, a philosophy of the pursuit of knowledge for its own interest is both right and proper. At the level of a sequence or province of knowledge, however, the dominant consideration is the importance of the particular research to the over-all knowledge. Failure to recognize these differences of orientation is a fruitful source of misunderstanding and mistrust. Their reconciliation is largely a matter of scientific management. In research, policy for the development of a particular sequence or province of knowledge, expresses itself not through prescription to individuals but by the informed selection of interests to be supported. Once that is decided then the more the individual is left to follow his own inspiration, the better.

Nevertheless, individual men tend inevitably to see the significance of their own activities largely in terms of their particular interests and, in seeking, as we all do, to justify themselves, to adopt ideas of the structure of scientific knowledge that will not minimize these. Thus those working towards the specialized extreme of the sequence of knowledge will naturally incline to see all other subjects in the sequence in terms of their contribution to the special interest in question. Equally those working towards the unspecialized extreme will tend to adopt concepts that stress the dependence of more specialized subjects on the development of unspecialized knowledge. Now that specialization of research has developed to the extent that it has, such differences in viewpoint at the individual level are practically inevitable and account in large part for the divergencies of opinion regarding the way scientific knowledge develops. If, however, the sequences of knowledge that we have considered in the biomedical field are any indication of the structure of scientic knowledge in general, then we might hope that consideration at the level of these rather than individual subjects might enable such differences to be resolved.

V

In our further approach to this problem of the over-all structure of scientific knowledge it would appear, therefore, that there are two considerations that we should keep in the forefront of our minds. The first is that we should be clear as to the level from which the problem is being viewed so that due allowance can be made for any bias that specialization of interest, or viewpoint, might introduce. The second is that the full intellectual significance of any particular subject is determined not by its own intrinsic properties but by the importance of its relation to the larger intellectual context of which it forms a part. If, therefore, we could identify the various sequences of context that go to make up scientific knowledge in general, we should be in a fair way to obtaining a comprehensive picture of its over-all structure.

In respect of biomedical knowledge, the various sequences of knowledge stretching from the specialized periphery to the more unspecialized central regions, and which collectively constitute the whole province of biomedical knowledge, appear to provide the intellectual continuum within which the individual subjects find their natural place and context. The question that now requires consideration is whether this is a peculiarity of that province or whether it is a general feature of scientific knowledge.

4

AGRICULTURE, MATERIALS AND ENERGY

I

IN THE PRECEDING PAGES we have considered examples of sequences of necessary knowledge within the biomedical field and the interrelationships between them that appear to weld these into a major province of natural knowledge. Because of my personal familiarity with knowledge in this particular field, I have felt reasonable confidence in so doing. But nowadays no man could hope to have equal familiarity with all other scientific fields and it is with very real diffidence that I now propose to examine these from the viewpoint that has been put forward. Nevertheless there is a reason why this is unavoidable. Although people may be prepared to concede that the concept put forward may apply to the biomedical field, they tend to shrink from contemplating its universality and to postpone consideration by suggesting that biomedical knowledge may be a special case. On the face of it, it seems unlikely that the structure of knowledge should be unique in any one part of the scientific field. As long, however, as we are prepared to entertain this possibility, so long shall we be inhibited from arriving at any general concept of the structure of scientific knowledge and the requirements for its continued development. It seems, therefore, that we can no longer avoid examining this problem, and, daunting as this is, we may derive some assistance in so doing from a point that has already been made.

In discussing the examples of the sequences of necessary knowledge that could be dissected out of the biomedical province, it was pointed out that these sequences were not only pragmatical but also historical in form. Later, when considering the development of a line of scientific enquiry, and the evolution of the subjects within it, the historical nature of these processes emerged by implication. It might well be, therefore, that the most promising approach to this problem is an historical one.[1] This I propose to take but, in doing so, I wish to insist upon one point. It would indeed be foolish to suggest that the early course of development of any natural knowledge was in any way the expression of a conscious plan. Obviously it was not. Rather it was an expression of the activity of individuals, motivated by need or curiosity as the case might be. Nevertheless, from what remained after all their failures, a kind of pattern upon which knowledge has developed does seem to emerge. It is this pattern, with its interplay between specialized and unspecialized knowledge, with which we are concerned.

Let us start by considering the province of knowledge that is nearest to the biomedical, that of agriculture.

II

It is evident that from the beginnings of agriculture, man must have been concerned with the varieties of edible plants, the quality of the soil, the supply of water and the succession of the seasons. Even at these early stages, however, a sequence of knowledge was already being established. As men passed from the stage of food gathering to that of cultivating, they changed over from dependence on chance to dependence on planning. They thus embarked on the sequence they have been pursuing ever since: the accumulation and systematization of the necessary knowledge in order to arrive at a basis for prediction.

What this has led to in agriculture can be seen by a glance at the sequences of knowledge (and research) that is now necessary to this end. In doing so, however, it is necessary to bear one thing in mind. In any given situation, varieties of plants and animals, soil, water, weather and the activities of man himself form one ecological system and, although it may be necessary for

[1] In this historical approach I have drawn heavily upon, and been substantially guided by, the book *Science in History* by J. D. Bernal (C. A. Watts & Co. Ltd, London, 1965), and I should like to put on record my indebtedness to this.

ease of consideration to break this down into its components, these only appear in their full significance in relation to the system as a whole.

Let us first consider edible plants. Out of the three hundred thousand or so plant species that are known, only some 3,000 have ever been used for food, only some 300 are widely grown for this purpose and we rely for over 90 per cent of our food on a mere dozen or so.[1] That such a small number of species can fill such a major role is an indication of the knowledge of selection, breeding and cultivation that has been brought to bear. From time immemorial, men have been selecting plants on the basis of their relative food value either to themselves or to their domesticated animals. Early they must have discovered that crossing of natural variants might produce offspring of increased value so that, over the centuries, there accumulated a traditional lore on plant and animal breeding. It required the genius of Mendel, however, to introduce the elements of precision and theory into this and so to lay the foundations of the modern subject of genetics. From this clarification, knowledge has developed apace. The identification of chromosomes as the vehicle of genes, demonstration that mutagenic agents such as radiations and certain chemicals would modify these, have led to the recognition of a sequence of the knowledge necessary for effective work that stretches from the informed empiricism of modern trials of cross-breeding, through the theory of genetics to cell biology, the biochemistry of nucleic acids and research at the molecular level. But it would be quite erroneous to attribute the recent substantial progress in producing improved varieties of useful plants or animals entirely to the influence of developing genetic theory. Certainly this has played a part but most of the successes have been achieved by much more rough and ready methods. The characters of weight, yield and so on, depend upon genetic factors of such complexity that theory has not yet fully taken the measure of them.[2] The further development of genetic knowledge thus

[1] J. G. Harrar: *Strategy towards the Conquest of Hunger*, p. 100. The Rockefeller Foundation, New York, 1967.
[2] J. D. Bernal: *loc. cit.*, p. 701.

depends in no small degree on the continued stimulation of the more theoretical studies by the specialized knowledge gained in mission-orientated situations flowing down to them through the continuum of knowledge of which they are all a part.

Again, consider our understanding of soil fertility. Bitter experience must early have convinced men that there were good lands and bad lands, and later, that it was possible to exhaust the good and turn it into bad. On the basis of such experience, the practical art of husbandry, with its accumulating traditions of irrigation and fertilization, fallow periods and crop rotation, would slowly develop. But look at the sequence of necessary knowledge that has now grown from this. It stretches through agro-chemistry, physical chemistry, microbiology, chemical microbiology, microhydrodynamics, and, although it may not yet have reached the molecular level, it is clearly tending rapidly in that direction. Again the two-way flow of knowledge along the continuum is evident. And, looking to the future, almost certainly the problem of nitrogen fixation alone will, before it is solved, open as many chapters at all levels of our knowledge as that of photo-synthesis has already done.

The essential similarity in form of these examples to those cited in the province of biomedical knowledge is apparent. Further, as one approaches the more unspecialized levels, the different sequences of agricultural knowledge find increasingly more common ground not only within their own province but in that of biomedicine as well. Knowledge of virology stimulated by the problems of the virus diseases of plants flows right across to the knowledge stimulated by the virological problems of human disease. Knowledge of the cell biology of plants fuses with similar knowledge of the cells of animals. All this is not surprising. But what when we move from the provinces of knowledge related to biology to the provinces related to the physical sciences?

III

Man's need for materials necessarily led him to search for their sources. Early in his development he must have started to distinguish between the different kinds of rocks, and the seams within these, in the light of the purpose that provided his incentive. Mining was the natural sequel to collecting and primitive earth grubbing. The empirical knowledge and lore thus accumulated in these crude activities provided the basis for development of the earth sciences.

With the depletion of the more easily accessible sources of mineral wealth, increasingly sophisticated physical techniques of exploration and increased understanding of the structure of the earth's crust became necessary. The span of necessary knowledges lengthened correspondingly and these, in their turn, were stimulated by their new experience to develop and (much like the development of biology had led to the development of biochemistry and broadened the experience of organic chemistry) a new discipline of geophysics came into existence.

With the development, both intellectual and utilitarian, of these studies on the structure of the earth, these sequences of knowledge came increasingly together with the sequences that had started in the need to chart the oceans and which were themselves increasing in sophistication. Again, consequent on the development of air travel, the need for knowledge of wind currents, visibilities and air temperatures demanded degrees of precision hitherto quite beyond earthbound human requirements. Knowledge of the earth's atmosphere expanded correspondingly. Meteorology, like geology and oceanography, became increasingly sophisticated and, in so doing, not only extended the sequences of knowledge of which they had need but provided stimulation to the further development of the subjects within these. The course of the development in the understanding of our physical environment, and the form taken by development in the provinces of biomedical and agricultural knowledge, have evident resemblances.

IV

Let us now move to man's endeavour to understand the materials of his environment, an endeavour that has led to the body of knowledge that we call chemistry. Broadly speaking, this appears to have started from three main needs: utensils and fabrics, metals and drugs.

From a surprisingly early period in his history, man appears to have been imbued with the desire for decoration and adornment. So much is this so that the increasing sophistication of the designs and colours on pottery is a useful pointer to the development of primitive man. Naturally occurring pigments, their behaviour under different physical conditions, their interaction, their extraction in purer form from the natural state in which they were found, must have been an early and prized preoccupation. Alongside this the need for metal—iron, gold, silver—and the separation of these from crude ores, their purification and manipulation, was leading to a convergent line of interest which, in its early stages at least, was not the less urgent because of the mystical incentives with which it was endowed by alchemy. And alongside this again was the line of search for better materials with which to treat illness, although this was somewhat further removed because of its primary concern with organic rather than inorganic materials. The outcome of centuries of such activities was the accumulation of a vast mass of empirical knowledge within which the beginnings of order slowly began to emerge in the identification of different materials, their properties and interactions and their categorization by similarities. This was the position by the mid-17th century in Europe. Thereafter, with the extension of the quantitative approach, the way became clear for the emergence of the idea of proportionate combination, the atomic theory and all the impressive structure of chemical theory that followed this. Understanding of organic chemistry—'the chemistry of carbon compounds'—followed more slowly until the concept that particular arrangements of the same elements have particular properties was introduced by Liebig and Wohler, and the idea of molecular shape was suggested by Pasteur's work.

It is evident that in its ancestry chemistry, dealing as it does with the properties and manufacture of materials, was closely bound up with industrial need. Even when chemical enquiries had developed to the levels at which theory could be formulated, this persisted. One of the most significant events in this respect was the discovery by Perkin in 1856, whilst searching for a substitute for quinine, of the first synthetic aniline dye.[1] After some time, this captured interest and the chemical industry came into being in its wake. With such an antecedent, it was inevitable from the start that research should be an integral part of this industry. But it was research that at one extreme was highly sensitive to its mission. The result was that a diversity of detailed experience poured in and posed a flood of new questions to chemical theory at the more unspecialized levels. From these beginnings the use of chemical concepts, knowledge and methods spread to field after field as these developed to the stages at which they required chemistry for the analysis of their own problems. Indeed, so vast is the spread of chemistry across several of the major provinces of natural knowledge, that for one who is no expert it is only possible to select highly specialized points for illustration of the present position.

An example of the width and range of research now flowing from scientifically based practical activities was given by Mees, Vice-President of the Eastman Kodak Company, whilst holding the Hitchcock Professorship at the University of California in 1946.[2] He pointed out that photography was discovered before there was any understanding of its theory and that even now the reason for the properties of photographic emulsions or the effect of different dyes in altering the properties of the light-sensitive chemicals in these were largely unknown. And he then goes on to indicate the wide range and depth of research that an industrial corporation must today carry out in the fields of analytical chemistry, physical chemistry, optics and physical theory.[3] All these not only draw upon deeper levels of unspecialized knowledge, but themselves define problems for such knowledge to solve, and, in the process, to extend the development of its own theory.

Jewkes, Sawers and Stillerman in their book *The Sources of Invention* give numerous examples of situations of similar significance.

[1] Bernal: *loc. cit.*, p. 458.
[2] C. E. K. Mees: *The Path of Science*, p. 43. John Wiley & Sons, New York, 1946. [3] Mees: *loc. cit.*, p. 208.

AGRICULTURE, MATERIALS AND ENERGY

The discovery of nylon provides another variant. Carrother's work on the formation and structure of high molecular weight polymers was the starting-point: the chance observation of Hill that a particular polymer, when molten, could be drawn out into a fine thread and that the resulting fibre, when cold, could be further drawn out and (an entirely new phenomenon) its strength and elasticity thereby increased, suggested its practical relevance. Clearly, the possibility of artificial fibres, with all that this implied for the textile industry, had become a live issue. But the fibre from this particular polymer proved useless for these purposes and efforts were redoubled to find one with the required specifications. Under the stimulus of this quest, and aided by the experience already gained, the theory of fibre structure developed rapidly and eventually a polyamide compound was produced from which a strong, tough, water and temperature resistant fibre could be drawn. The advance of knowledge and theory in the several subjects necessary to these operations has now led to such increased understanding as to bring us within range of producing artificial fibres to specification. Here we have an example of a sequence starting at a relatively deep level and developing both by stretching up to more specialized levels and delving down to deeper unspecialized.[1]

The solution of the 'anti-knock' problem in motor-car engines provides an example of a sequence starting at a completely specialized level. The problem of 'knock' in such engines, and consequent loss of efficiency, had been known for a long time. The first step was an accurate investigation of the actual phenomenon, the elimination of mechanical factors as the agent, and the identification of fuel as the cause. Attempts were then made to 'doctor' the fuel and, at first, various dyes were used. A chance observation then turned attention to certain metals and these were immediately found to be more effective. Further experience related the efficacy of the metal in this respect to its position in the periodic table and, on this basis, lead appeared to be the metal of choice. But no-one knew how to make a preparation of

[1] J. Jewkes, D. Sawers and R. Stillerman: *The Sources of Invention*, p. 334. Macmillan and Co. Ltd, London, 1955.

lead that was soluble in petrol. The centre of effort thus moved down the sequence to chemistry, and after intensive research at the purely chemical level tetra-ethyl lead was produced.[1]

As a final illustration in this field we can take an example from the field of fine chemicals: the discovery of cortisone. The story started with the validation by Hench of the clinical impression that the intractable disease rheumatoid arthritis might remit during pregnancy and during an attack of liver disease. Searching his mind for a common factor in these two widely different conditions, it occurred to Hench that in pregnancy there was a greatly increased production of steroid hormones by the relevant endocrine glands while in liver disease, although there was no increase in production, there was a decreased rate of destruction of these compounds. The net result was, therefore, that in both conditions an increased concentration of steroid hormones developed in the body fluids, and the obvious indication was to test the effect of giving different steroids on the course of rheumatoid arthritis. Fortunately in the course of investigating the function of the endocrine glands, many such compounds had been isolated, and one of these, cortisone, proved to have the desired effect. But cortisone was extremely difficult to prepare and the known methods yielded amounts that were quite insufficient. A synthesis with high yield and starting from materials that were not prohibitively expensive was clearly the only answer. An intensive search then developed over the whole globe for a naturally occurring starting material with a chemical structure that was already developed beyond the stage that presented, at that time, an insuperable difficulty in large-scale synthesis. Eventually such a material was found among the waste product of the sisal plant. In this sequence of clinical research, pathology, biochemistry, and chemistry, we have a collection of subjects that were all necessary to the pioneering effort. Since then in the search for understanding, all these subjects have developed and further subjects like cellular biology and immunology have been drawn in. In this succession each subject provides questions for the next, yet each develops its characteristic knowledge; and not the least

[1] Jewkes et al., loc. cit., p. 392.

of these subjects is chemistry that, under the stimulation of the task, has developed its theoretical knowledge of the structures and interactions of steroid compounds.

These examples are but fragments and this is necessarily so. As has been pointed out, the study of the nature of materials derived, historically, from three main needs—fabrics and utensils, metals, and drugs—and it was on the basis of the data and problems thrown up by these that the discipline that we call chemistry was built. Now it has not only acquired many more sources of data, such as fuel technology, biology and so on, but it has itself, as it developed, begun to produce artificial materials which, under trial in specialized conditions, have themselves become sources of data for the further development of chemical knowledge. Consideration which concentrates on a single discipline like chemistry is thus different from consideration in respect of a sequence of necessary knowledges. In chemistry proper we are dealing not with the whole span of a sequence, but only with that part of it that is concerned with the composition of its material. This is an important distinction. To clarify it, consider the situation in respect of a cell. Knowledge of the nature of the materials within the cell and their interactions is essential, but in itself this will not give us a complete understanding of the cell as a functioning unit, still less of the cell as a functioning component of a multicellular organism. For such understanding, the whole sequence of relevant knowledge is necessary. In considering a particular subject (however well developed and however widely ranging this is) we are in consequence restricted to those parts of sequences with which it is concerned. Examples to illustrate the place of an individual subject like chemistry in the structure of natural knowledge must, therefore, from the nature of the case, appear to be fragmentary and of restricted significance.

VI

Let us now turn to the most highly developed of all scientific knowledges, the physical sciences, and the most developed of these, that concerned with energy.

From very early days men have sought to supplement their own strength or to devise means by which they could exert it more effectively. The sail, the windmill and the water-wheel were early attempts to harness the free play of natural forces. The lever and the inclined plane even earlier devices by man to increase the effect of his own exertions. It was not, however, until the invention of gunpowder that men were faced with the possibility of liberating force on their own volition and directing it to their own ends.

The requirements of these endeavours were the same in each case and have remained the same to this day. They are a source of energy and a machine to harness this. These requirements must be met whether the source of energy is man's own physical strength, the force of the wind, the molecular energy stored in fuels, or atomic energy. Thus our understanding of energy has marched hand in hand with our ability to invent machines and, in general, the pattern of development in our knowledge of physical energy has followed the same course: identification of a source of energy, invention, development of theory, improvement of invention in the light of increased theoretical understanding. As an example, we may take the train of development that led up to the perfection of the steam engine, a machine for controllably translating heat into mechanical energy.

The invention of the piston pump is lost in the mists of antiquity. It was not, however, until the 17th century that Torricelli found the correct explanation of the familiar observation that a suction pump would not raise water by more than some thirty-two feet, namely that a column of water of that height corresponded to the weight of the atmosphere at sea level.[1] Clearly, it followed from this that if gas could be removed, or its volume reduced, on one side of the piston, the pressure of air on the other side would force the piston into the cylinder; and then, if gas were reintroduced into the cylinder, the piston would slide back. Interestingly enough, the first attempt to take advantage of this knowledge, by Papin, relied on the cooling of hot gases under the piston, such

---

[1] Phillip Lenard: *Great Men of Science*, p. 49. English translation by H. S. Hatfield. G. Bell & Sons, Ltd, London, 1933.

gases having been produced by the explosion of gunpowder.[1] It was a short step from this to introducing steam and then condensing it. Essentially, such steam pumps (or engines) relied not on the propulsive power of steam, but on the production of a 'vacuum' for obtaining their effects, and, as the cylinder had to be cooled after each stroke to obtain this, their inefficiency and consumption of fuel was great. It was here that the situation was revolutionized by Watt with his invention of the separate condenser. In this, Watt was undoubtedly influenced by his chief, Joseph Black, the discoverer of latent heat, which discovery pointed the way to economizing on the great loss of heat. He was further aided by the advance of engineering technology which made possible the construction of high-pressure boilers and the utilization of the propellant force of steam pressure. The outcome was a practical economic steam engine.[2] But, invention had far outrun theory. It was left to Sadi Carnot, some thirty years later, to elucidate the problem of the transfer of energy from heat to mechanical efficiency in the steam engine and, in so doing, to lay the foundation of theoretical thermodynamics.[3] In turn, this theory led to the further improvement of steam engines and later provided the background against which the internal combustion engine was devised.

The sequence here well illustrates the pattern. The invention of the suction pump. The theoretical understanding of the nature of a vacuum. The invention of a machine which made possible the production of a 'vacuum' by the condensation of steam. The independent line of investigation that led to the theoretical concept of latent heat and the recognition that this provided the explanation of the inefficiency of existing steam engines. The invention of an effective engine, the study of the basis of its efficiency and the subsequent foundation of a new branch of physical theory.

As the next illustration, take the development of our knowledge of electricity. In emphasis this differs from the previous one in that the energy with which one is here dealing does not exist in

[1] Phillip Lenard: *loc. cit.*, p. 118. [2] Lenard: *loc. cit.*, p. 130.
[3] Lenard: *loc. cit.*, p. 231.

THE STRUCTURE OF SCIENTIFIC KNOWLEDGE

nature in a form that can readily be handled but has to be produced from other more tractable sources. As a consequence, invention as a source of intellectual stimulation does not bulk large until the later stage of the account.

Like the suction pump, the origin of the mariner's compass is lost in the distant past, but, with the increase in navigation in the later middle ages, the improvement and understanding of this invention became of increasing importance. The systematic approach to knowledge started at the end of the 16th century, with the publication by Robert Norman, a compass-maker and sailor, of a practical treatise on the use of the instrument, and the great theoretical work of the physician, William Gilbert, that stemmed from this.[1] Gilbert showed that the behaviour of the compass could be explained by the earth being a great magnet.[2] He was thus led to the concept of forces, imperceptible to the senses, that were attractive, and he even went so far as to suggest that it was such forces that held the planets in their places.[3] Further, he investigated the attractive property conferred by friction on substances like amber, and correctly appreciated its close similarity to that of magnets. A century after Gilbert, Gray[4] showed that the 'imponderable fluid' responsible for electrical phenomena would flow along a damp string. Electric machines for producing powerful sources of this fluid were invented (largely for amusement as toys), and condensers to store this followed. Fifty years after that, Benjamin Franklin saw that there were not two forms of electricity, but only one; bodies becoming positively charged if the 'imponderable fluid' were added and negatively if abstracted.[5] He also appreciated the nature of lightning and, with a characteristic excursion into the practical, devised the lightning conductor. Twenty-five years later still, Coulomb, while seeking further to improve the marine compass, established that the forces between magnetic poles, as well as those between opposite charges of electricity, obeyed the same laws as gravity.[6] Another

---

[1] E. Zilsel in *Origins of the Scientific Revolution*, edited by H. F. Kearney, p. 95. Longmans, London, 1964.
[2] C. E. K. Mees: *loc. cit.*, p. 102.
[3] J. D. Bernal: *loc. cit.*, p. 301.
[4] J. D. Bernal: *loc. cit.*, p. 432.
[5] J. D. Bernal: *loc. cit.*, p. 433.
[6] J. D. Bernal: *loc. cit.*, p. 434.

fifty years had to pass, however, before Oersted observed that a compass needle was deflected by an electric current flowing in proximity to it.[1] Then, hard on the heels of this, came Faraday's crucial discovery that, if a magnet were moved near an electric conductor, an electric current arose. He thus established that it was possible to generate electric current by mechanical action and, equally important, mechanical action by electric current.[2] Thereby the whole course of development was set in train that led to control of an enormous and convenient new source of energy.

Whilst all this was happening, other sequences were developing and converging. The dependence of man upon sight had early directed attention to the eye and to speculation on the nature of light. Lenses were certainly known to the Arabian physicians and, although these were at first of crystal, the trades of lens-grinder and spectacle-maker became well established when clear glass became available. The juxtaposition of two lenses to form a telescope is traditionally due to an accident. It came sufficiently early, however, for Galileo to see the moons of Saturn (and thus vindicate the Copernican system) and for Kepler to write a book on its theory. Discovery of the phenomenon of refraction followed quickly, to be followed in its turn by Newton's great work on the spectrum showing that white light was made up of rays of different refrangibility. In consequence of this discovery, Newton came to believe that it would be impossible to make perfect lenses, that is lenses free from chromatic aberration.[3,4] But the need for such lenses was insistent and in the next century, the lens-maker Fraunhofer[5] achieved success and discovered his famous lens which enabled defined kinds of light to be isolated.

From the beginning the immense speed of light had been evident at every dawn. That it was not infinite but of the order of 200,000 miles a second was shown before the end of the 17th century. Throughout these developments speculation on the nature of light was inevitable, and during the 18th century Newton's corpuscular theory and Huygens's wave theory each

[1] J. D. Bernal: *loc. cit.*, p. 437.
[2] J. D. Bernal: *loc. cit.*, p. 438.
[3] J. D. Bernal: *loc. cit.*, p. 327.
[4] C. E. K. Mees: *loc. cit.*, p. 98.
[5] P. Lenard: *loc. cit.*, p. 198.

commanded support. Early in the next century, however, considerations of the phenomena of polarization and interference centred views on the wave hypothesis. Now there were three imponderable fluids that passed through, apparently, empty space. It was left to the genius of Clerk Maxwell to unite these in one theory.[1, 2] This, taken in conjunction with the emerging laws of thermodynamics, pointed to a hitherto undreamed-of unity in the nature of energy.

This synthesis had great predictive significance. It indicated that electromagnetic oscillations should produce waves similar to those of light but of lower frequency, and later in the century these were, in fact, produced by Hertz and wireless telegraphy became a possibility. This has brought us almost to our day and the subsequent development is instructive.

Signalling by light was an ancient way of conveying information. Signalling by electric current along wires was established. The possibility of using the new electromagnetic waves for the same purpose was bound to suggest itself. But current physical theory discounted this possibility. Such waves, it was predicted, would fly off at a tangent from the surface of the globe and lose themselves in the void. Fortunately, Marconi remained impervious to theory and tried. Outstanding success was the result.[3] But this totally unexpected success had now to be explained. Clearly, as suggested by Heaviside, there must be something like a mantle round the earth that reflected the rays and kept them to the globe's surface. That this was indeed so was shown by Appleton. Further, the property of reflectibility was put to practical use (on the model of echo-sounding with sound waves) as a means of locating objects at a distance—and radar came into existence. Further technical development of this has now become the basis of radio-astronomy.

But the most dramatic manifestation of this developing knowledge of electricity had yet to come. To reach it, we must first go back in time.

Lightning had always been a puzzle. Air was a non-conductor,

[1] J. D. Bernal: *loc. cit.*, p. 405.   [2] P. Lenard: *loc. cit.*, p. 339.
[3] J. D. Bernal: *loc. cit.*, p. 552.

yet in certain circumstances electricity would jump through it. This was the basis of carbon arc lighting. Whilst studying the passage of electricity through gases, Faraday had noted that a glow appeared in an almost completely evacuated tube when a current was passed. Later, Hittorf and then Crookes[1] showed that the glow was due to a stream of charged particles that could be deviated by a magnet. Then Roentgen discovered X-rays apparently almost by accident. From the end of the evacuated tube, where the glowing stream impinged, came invisible rays that fogged photographic plates, traversed the body and made certain chemicals fluoresce.[2] Interest was widespread. Naturally, phosphorescent substances had always attracted curiosity and, stimulated by the new discovery, Becquerel examined a specimen of uranium nitrate and found that it did indeed emit rays similar to those produced by Roentgen.[3] Radioactivity had been discovered and the way to explore the structure of the atom opened. The brilliant story that followed has been told many times. By the turn of the century Rutherford had shown that radium was giving off helium atoms and changing into radon. Other radioactive substances were undergoing further mutations and from Einstein's work, the conversion of mass into energy with the possibility of immense liberation of energy, was appreciated. Although Rutherford refused to believe that this tremendous new source of energy could be put to useful work, the day of atomic power had dawned. Under the threats of war it developed with unprecedented speed. By 1940 it had been proved that mass and energy were equivalent, that the neutrons initiating fission of uranium reproduced themselves in the process and that, therefore, a multiplying chain reaction might occur. It was well said, however, that the 'gap between producing a controlled chain reaction and using it as a large-scale power source or explosive, is comparable to the gap between the discovery of fire and the manufacture of the steam locomotive'. But by 1945 the problem of producing an atomic bomb had been solved in essence, with the results that we all know. In the process, obstacles of the most

---

[1] P. Lenard: *loc. cit.*, p. 344. [2] J. D. Bernal: *loc. cit.*, p. 521.
[3] J. D. Bernal: *loc. cit.*, p. 523.

daunting and most varied types were surmounted. To give one example: the important isotope in natural uranium for the purposes is uranium 235, which occurs as one part to 140 of uranium 238. Being chemically identical, the separation of these two isotopes was impossible by chemical means. A completely new element, plutonium, was therefore made from uranium 238 which, being chemically different, could be separated.[1]

VII

As a last example, let us take transport. The essential requirements here are a source of motive power and a suitable vehicle. The earliest problem would almost certainly be that of crossing water. The raft and the boat were the vehicles and oars or sails the devices for using power until the invention of the steam engine. But the vehicles had to be adapted to their purposes and the endeavour to improve them led to the theory of design for the hull and screw and the emergence of fluid dynamics.

Let us jump now to the problem of flight. Attempts to fly have, despite the discouragement of danger, been made by a few hardy men over the centuries. Once they had abandoned the attempt to imitate birds and turned to the glider, success was near. With the invention of the light internal combustion engine as a source of motive power, it was assured. But as in the case of the steam engine, trial and error succeeded before theory was devised. The result was a tremendous stimulation of physical theorizing and the development of the new subject of aerodynamics. This in its turn led to the further development of aeroplanes and had wide repercussions in engineering science.[2] The development of rockets followed a similar course. Rockets as fireworks and as an unfavoured alternative to cannon have been known for centuries, but the jet as a source of motive power lingered until, under the exigencies of war, it was developed in the last three decades. The rocket as a vehicle, even more than the plane, owes its inception

---

[1] H. D. Smyth: *A General Account of the Development of Methods of using Atomic Energy for Military Purposes.* U.S. Army publication, Washington, 1945.   [2] J. D. Bernal: *loc. cit.*, pp. 579-81.

to the trials and errors of amateurs.[1] Now it is a reality and the problems that outer space has posed to theory are only at their beginning.

VIII

It would be possible to continue multiplying these examples. One could follow the train of knowledge that developed from the interaction of need and understanding into the serious study of astronomy; or that stimulated by the ballistic problems following the invention of cannon that led to the understanding of motion, force and inertia. But for our immediate purposes sufficient have been given for us to take stock of the situation and to consider whether the sequences of necessary knowledge, such as those distinguished in the biomedical province, are a feature of scientific knowledge in general and, if so, to form some estimate of their significance.

It would seem that, historically speaking, the train of scientific development has been from the particular towards the successively more general. This is evident in regard to the provinces of knowledge concerned with biomedicine, agriculture and the earth sciences. It seems also evident in respect of at least the early history of development of knowledge regarding materials and physical energy. Clearly in all these provinces investigations started with particular problems and, in the endeavour to understand these, enquiry progressed to successively deeper levels where knowledge became progressively more capable of generalization. Thus, in each province, the process of development was marked by particular knowledge coming down from, and more generalized knowledge coming up to, the periphery.

But are these sequences of development, with their elaboration of a succession of knowledges, anything more than an historical statement? Have they any present or continuing significance? I think that we have good grounds for concluding that they indeed have.

[1] J. Jewkes, D. Sawers and R. Stillerman: *The Sources of Invention*, p. 355. Macmillan, 1958.

It is accepted, almost without question today, that scientific knowledge can be categorized as basic, applied or developmental and, although we shall see later that this classification is open to serious objection, this at least expresses the tacit recognition that the different categories of scientific knowledge do not exist in isolation but rather in a meaningful relation to each other. Traditionally, the applied sciences are regarded as being built on the basic and the developmental on the applied. In effect, therefore, it is already generally accepted that there is a sequence in the arrangement of knowledges and further that the continued development of such knowledge depends on the existence of this.

Yet, although the concepts of sequence based either on the traditional classification of scientific knowledge, or derived from historical considerations of the way such knowledge has developed, agree in regarding their respective sequences as essentially significant rather than expedient, there is an important difference between them. It relates to the way in which scientific knowledge advances. On the basis of the traditional classification, the emphasis in the sequence of development is from the general (or basic) to the particular (or applied). On the basis of historical considerations, however, the emphasis is rather from the particular to the general. On the face of it there is, therefore, something of a contradiction between the significance attached to the concepts of sequence derived from these two different sets of considerations. The question is can these be reconciled?

The solution of the dilemma was put in a single sentence by Herbert Spencer more than a century ago: 'A more general science as much owes its progress to the presentation of new problems by a more special science, as a more special science owes its progress to the solutions that the more general science is thus led to attempt.'[1]

In principle, this view can hardly be gainsaid. Accordingly, it would appear justified to regard the sequences and provinces of natural knowledge we have sought to distinguish as indicating not

---

[1] H. Spencer: *The Genesis of Science*. Published 1854, republished in *Essays Scientific, Political and Speculative*, Vol. 2, p. 71. Williams and Norgate, London, 1891.

only the way that scientific knowledge has developed but also, and perhaps more importantly, as expressing the living and operative relationship between the individual subjects of scientific knowledge on which their future development and larger significance depend. It is on this basis, therefore, that I propose to approach the problem of the structure of such knowledge.

* 5 *
THE STRUCTURE OF SCIENTIFIC KNOWLEDGE

I

IT WILL BE CONVENIENT, at this stage, to summarize the main features of the concept that has been emerging from the previous discussion of which the unit is the sequence of necessary knowledges. The characteristics of such sequences are the following.

Each of these stretches from more or less specialized knowledge at one end, through progressively less specialized, to more or less unspecialized at the other. Intellectually speaking, each such sequence forms one continuum of thought. All the knowledges in it are necessary to the over-all intellectual synthesis, although the relative contributions of each to this may vary from sequence to sequence. Forming one single continuum, the component subjects of knowledge within a sequence are each in their appropriate intellectual context. Each is linked by directly relevant interests to the subjects on either side of it and it is through such linkage that the integrity of the sequence as a whole arises and individual subjects achieve their larger significance.

As sequences develop to progressively deeper levels, however, they tend to converge and even to coalesce. Two things may then happen. First, the developing sequence may make contact with a subject that has already come to a high state of development in other sequences and a new subject may come into existence. It was in this way that biochemistry emerged when developing biological knowledge made contact with chemistry. The second is that, as sequences converge and come to have similar kinds of data increasingly in common, groupings (or strata) of similar data running across many sequences begin to appear at the subject level, and, when these groupings become sufficiently large, it becomes possible to study the common features that underlie their similarities. It was in this way that unspecialized chemical knowledge emerged from the mass of data that came down to it from the numerous sequences that supplied data on materials and the generalizations of chemical theory became possible. But, within the intellectual continua that characterize sequences, the flow of knowledge is in both directions and it is on this dynamic interplay between specialized and less specialized (or unspecialized) knowledge that future development, not only of the sequence in question, but of the component subjects depends.

These are the basic considerations in the concept that I want to put forward. In themselves, however, intellectual sequences are only the units from which the over-all structure of knowledge is built up. Beyond these there are natural groupings of sequences. Thus the several sequences of biomedicine, of bioagriculture, of materials, of energy, have natural affinities with each other which bring them together into these larger groupings. It is these that I have called provinces of natural knowledge. Each of these has its peripheral frontier of specialized problems and each develops towards a common centre where knowledge becomes increasingly unspecialized.

This is the basis of the concept that I propose. It is significantly different from that to which we are traditionally accustomed.

II

When we have occasion to talk, or even to think, of the structure of knowledge we are preconditioned to use the analogy of a tree. Like all analogies, that of the tree of knowledge derives from previous and quite different experience. Although in western civilization it was first used explicitly in a theological context, the unitary concept of knowledge to which it gives expression was equally consistent with the trend of thought in classical philosophy. When, therefore, in the 17th century the accumulation of natural knowledge raised increasingly the question of its systematization, it would have been asking too much of men to expect them to shed their previous preconceptions and consider things afresh. As Bacon himself somewhat ruefully remarked: 'Even to deliver and explain what I bring forward is no easy matter, for things in themselves new will yet be apprehended with reference to what is old.'[1] It was not only that men had before them the fate that had befallen Giordano Bruno, Servitus, Galileo and others who had had the temerity to arouse the suspicion of the established order, but in the climate of belief then prevailing they had every personal inducement to shrink from unsettling their own peace of mind. The unquestioned monotheistic belief of those times and the identification of this with wisdom, implied the essential unity of all knowledge. It was no wonder then that Bacon should see as the basis of his endeavour the elucidation of what he called a *philosophia prima* and that he should have seen this as the trunk of the tree of knowledge. 'But because the distributions and partitions of knowledge are not like several lines that meet in one angle and so touch but in a point; but are like branches of a tree that meet in a stem, which hath a dimension and quantity of entireness and continuance, before it comes to discontinue and break itself into arms and boughs: therefore it is good, before we enter into the former distribution to erect and constitute one universal science by the name of *philosophia prima*, primitive or summary philosophy, as the main and common way before we come where the ways part and divide themselves.'[2] Thus, at the impressionable outset of modern scientific growth, men's traditional belief in the unity of knowledge, with the tree as the proper analogy for its structure and development, was endorsed by the leading protagonist of the new enlightenment.

Since that time our knowledge of natural phenomena has expanded enormously and it is only natural for us to assume that our thinking about scientific understanding must have kept pace with this. But analogies once enshrined in attitudes of mind die hard, and Herbert Spencer's views are of interest in this connexion.

---

[1] Bacon: *The New Organon*, Bk 1, *Aphorisms*, p. 59. Liberal Arts Press, New York, 1960.
[2] Bacon: *Of the Advancement of Learning*, p. 100. World Classics Series, Oxford.

THE STRUCTURE OF SCIENTIFIC KNOWLEDGE

Writing in 1854 under the title of 'The Genesis of Science' he was primarily concerned to refute Auguste Comte's view that the different parts of scientific knowledge were like 'une echelle encyclopedique'. To this end most of his long essay is concerned with the development of the sciences from the simpler to the complex and he says in his conclusion: 'Scientific advance is as much from the special to the general as from the general to the special.'[1]

At this time, however, Spencer does not indicate his ideas on the structure of scientific knowledge beyond a passing statement that these are in harmony with the view that 'the sciences are as branches of one trunk'.

Then writing in 1869, to close fifteen years of controversy with Comte, he constructed in his second essay, entitled: 'The Classification of the Sciences', an even more developed analogy of the traditional tree of knowledge. After distinguishing three main kinds of science, the abstract, the abstract-concrete and the concrete, he continues: 'Their relations can thus only be truly shown by branches diverging from a common root.'[2] Thus, despite the evidence he had collected in his previous essay, Spencer passes unquestioned the validity of the traditional analogy of knowledge and its implications.

Nowadays, when we are more averse to expressing ourselves in terms of natural analogies, we are in some danger of concealing from ourselves the influence such models have on our conceptual thinking. Yet we have only to recollect how frequently we use such phrases as 'branches of learning' or 'roots of knowledge', or use the word 'science' in the singular, to suspect that the analogy of the tree of knowledge is still an influential factor in our thought.[3] Yet in my view this traditional model is entirely mistaken. To it I wish to oppose one that is diametrically opposite.

---

[1] H. Spencer: *Essays Scientific, Political and Speculative*, Vol. 2, p. 71. Williams and Norgate, London, 1891.
[2] H. Spencer: *loc. cit.*, p. 94.
[3] 'For men believe that their reason governs words; but it is also true that words react on understanding; and this it is that has rendered philosophy and the sciences sophistical and inactive.' Bacon: *The New Organon*, Bk 1, *Aphorisms*, p. 59.

The model that I would put forward is that of a vast globe of primitive ignorance around the periphery of which there are a whole series of problems prompting men to seek knowledge. Among these are the series concerned with health and disease, that concerned with food, that concerned with materials, that related to structures, that related to energy, that related to communal living, and so on. From all these points of departure men have now begun to penetrate towards the centre. The different series of problems at the periphery mark the frontiers of the different provinces of natural knowledge, the penetrations the development of these towards the centre. Not all such penetrations have been developed to the same depth, but many are already coming together. As progress delves ever deeper it departs increasingly from the specialized problems from which it took its start. But the information obtained is constantly passing up to the surface and more specialized information being returned. So the intellectual unity of the endeavour is preserved. And, in the deeper less specialized levels the different provinces uncover strata of similar kinds of data that, on development, reveal wide-ranging systems of more unspecialized knowledge.

This, the model of a globe of ignorance with areas on the periphery from which enquiries are being driven in towards the centre, is the analogy I wish to put forward for the structure of scientific knowledge in opposition to the traditional analogy of the tree of knowledge. It is, I believe, consistent with the historical development of the different sciences and the context of intellectual relationships within which the different subjects find themselves. It further provides an indication of the way that knowledge might be expected to develop in the future that is not, unlike the traditional analogy, inconsistent in essential respects with previous experience.

We have thus a choice between two diametrically opposed analogies. On the one hand we have the traditional one of the tree of knowledge with a common stem from which the different branches develop; on the other that of a huge globe of ignorance from separate places on the surface of which enquiries are being driven in towards the centre. But, it may be asked, does this matter? After all, we know that these are only analogies and, surely, to represent that the choice between them is in any way momentous would be vastly to exaggerate their importance. To a very considerable extent, however, men think in terms of conceptual models, and, according to the model they choose, they predetermine to a greater or less extent the assumptions on which they base their approach to future problems. The implications of any model that we adopt are, consequently, something that we cannot afford to neglect. I propose, therefore, now to look at the implications of the two analogies before us from the point of view of their bearing on the two most important questions that we must answer if we are to arrive at any concept of the structure of scientific knowledge. The first of these is the question of the unity of science; the second that of the categorization of the component subjects of scientific knowledge.

III

We can put the first question in one sentence: 'Is science a single entity or are there in fact many sciences?'

If we favour the analogy of the tree of knowledge with its common stem from which the different branches arise, we shall clearly be predisposed to accept the first alternative. If on the other hand we adopt the analogy of a globe of ignorance from the surface of which separate penetrations are being driven in, we shall equally clearly be predisposed to the second. What are the facts?

Historically there can be no question that the enquiries into natural phenomena that have led to the growth of knowledge we call scientific have started at many different points. Not only Spencer but Bacon himself recognized this when he traced 'experimental history' back to the different mechanical arts or a number of crafts and experiments which had not yet grown into an art properly so called.[1] There can equally be no question that, at this present time, there is little in common between several of

---

[1] 'History of arts and of nature as changed and altered by man, or experimental history, I divide into three. For it is derived either from the mechanical arts, or from a number of crafts and experiments which have not yet grown into an art properly so called. . . .' Bacon: *The New Organon*, p. 277.

the different parts of science. Astronomy, for example, has little if any common ground with biology. Cancer research and space research are, scientifically speaking, incommensurables. It might, however, be contended that as the different parts of science develop they come to have increasing ground in common and that, although many may not yet have developed to the stage of making contact, they will do so in due course and then they will form a unity. But, surely, this is to concede that the analogy of the tree with a common stem of knowledge from which all others develop is quite inappropriate.

Is it possible, however, that when all the parts of science are fully developed they will all come together in a common body of fundamental knowledge that will enable us to understand each and all? In the sense that the human body, plants and rocks may all be mapped in terms of their atomic structure this may be true, but, although this would be an intellectual *tour de force*, it is difficult to see how such knowledge could ever be the basis from which an understanding of these objects in the fullness of their existence could be derived.[1]

It was considerations of this kind, I believe, that led Karl Pearson to insist that 'The unity of science consists alone in its method, not in its material.'[2] Later, he elaborates what he means by this method: the classification of facts, the establishment of their mutual relation and sequence, and the construction by men whose imaginations have been so disciplined of rigorously verified conceptual models that subsume increasingly wide ranges of data. But if we accept, as I do, Pearson's view that the unity of science lies not in its material, but in its methodology, what becomes of the idea of a unity in scientific knowledge? Natural knowledge is knowledge of material things. Apart from this it has no meaning. We should expect, therefore, that there would be as many natural knowledges as there are groups of natural phenomena.[3]

[1] 'One of the strangest assumptions of modern biology is that knowledge of living man will automatically follow from so-called "fundamental" studies of the elementary structures and reactions of fragments derived from living things.' René Dubos: *Daedalus*, Winter, 1965, p. 242.

[2] K. Pearson: *The Grammar of Science*, p. 16. Everyman Library.

[3] Pearson himself divides scientific knowledge into concrete and abstract,

This is in effect what we find, and, although in the course of their further development these different natural knowledges will almost certainly come to have an increasing number of their component parts in common, these will still remain subsidiary to the over-all intellectual significance of the province in question.

It would seem, therefore, that the concept of science as a single entity has no foundation in fact. That it has been held so widely and for so long seems to be in no small measure due to the persisting influence of those philosophical and theological systems that have as a central tenet the ideal of a perfect knowledge or wisdom from which all else stems and find in the analogy of the tree an appropriate model for this. On the other hand, the facts suggest a concept of scientific knowledge that sees it as made up of a series of knowledges, each corresponding to a group of natural phenomena. For this, the analogy of a globe of ignorance into which penetrations are being made seems not inappropriate.

IV

The second question needing consideration is that of the categorization of scientific knowledge. On the basis of the traditional analogy one would expect that any such classification would be hierarchical; and this is, in fact, what one finds.

It is evident that the terms of the traditional classification 'basic' (or 'fundamental' or 'pure'), 'applied', and 'developmental', are relative. Like the terms big, bigger and biggest, each is a quality that can only be defined in terms of another and none can have any meaning apart from the situation in which it is used. It is also evident that these terms, being relative, imply sequence, and further that any such sequence is a self-contained hierarchical system without any independent point of reference outside itself. Certain consequences follow from the use of a system with this construction.

There being no independent criteria by which any one of

---

and the former into the subdivisions of organic and inorganic. Although not arranging these in any hierarchy he stresses the interactions between them. K. Pearson: *The Grammar of Science*, p. 320.

these terms can be defined, it is in consequence incumbent upon the individual user to decide for himself in each instance the point from which he will apply the system to any sequence of knowledges and thereby establish the status not only of the subject chosen, but also that of all others within the particular sequence. Thus, before a system of this kind can be put to use, a personal judgment of quality must be made. In principle, of course, this raises no insuperable objection to its use. If there is already close agreement as to what subjects merit description as 'basic', or 'applied' or 'developmental', the classifications produced by different individuals will correspond. If however there is no such agreement, but only a general kind of feeling surrounded by a broad penumbra of doubt, the use of such a system can only lead to confusion. That this latter is the situation in respect of scientific knowledge is suggested by the impatience that any request for a definition of these terms arouses.

In these circumstances, the hierarchical nature of the system (presupposing, as it does, a scale of value) becomes a serious weakness. It is, of course, theoretically possible to use an adjective of quality (although perhaps not of relative qualities) without committing oneself to an opinion as to whether the property described is important or unimportant. Yet the implications of common parlance are pervasive, and in ordinary usage the term 'basic' (and even more the term 'fundamental') carries an inescapable connotation of 'important'. It thus follows that the sequence of 'basic', 'applied' and 'developmental' comes inevitably to imply a descending order of importance and, in consequence, commits the user automatically to a presupposition of the relative value of the subjects under consideration for the development of scientific knowledge. But there are other and more serious objections.

Any sequence of relative terms that begins with the term 'basic' (or 'fundamental') inevitably implies that the essential flow of knowledge is in one direction: from 'basic' to 'applied' to 'developmental'. It thereby predisposes us to assume that this is the way scientific knowledge develops and to discount the significance of the contributions from the more specialized levels of

THE STRUCTURE OF SCIENTIFIC KNOWLEDGE

natural experience. Yet historically speaking this is patently an incomplete, if not misleading, representation of the case. One has only to recall the development of a subject like chemistry to see this. Historically, as we have seen, chemistry originated from the efforts of the dyer, the potter, the metalworker and the apothecary to understand the nature and reactions of the materials with which he was concerned, and it was from the development of these lines of enquiry and their increasing coalescence that the present impressive structure of chemical theory arose. Yet, if the traditional classification be accepted we should infer that the course of scientific development had been in the opposite direction. It could, of course, be objected that, although these contentions may be true from a historical point of view, it would be pushing matters to extremes to claim that today they represent anything like a complete account of the present situation. And that would be quite justified. Modern chemical knowledge and modern chemical theory by making it possible to synthesize not only naturally occurring materials but also materials that have never occurred in nature, have profoundly influenced the specialized levels from which chemistry derives and widely extended their range and efficiency. But it would be equally wrong to assume that the current of development had now been entirely reversed. The multiplicity and diversity of the requirements of modern society for different materials have enormously increased the sources of data from which to build chemical knowledge, and the vast expansion of activities to which such knowledge is relevant has made available a greatly increased range of situations in which chemical theory can be subjected to verification. If in the past the predominant current was from specialized to unspecialized knowledge, a more significant balance has now been struck. Herbert Spencer, as we have noted, put the matter clearly when, over a century ago, he wrote: 'A more general science as much owes its progress to the presentation of new problems by a more special science, as a more special science owes its progress to the solutions that the more general science is thus led to attempt.'[1]

In my opinion it is on such a dynamic interplay between

[1] H. Spencer: *Essays Scientific, Political and Speculative*, Vol. 2, p. 71.

specialized and unspecialized knowledge that the future development, not only of particular subjects but of scientific knowledge as a whole, substantially depends, and it is because the traditional classification of such knowledge obscures this that it is open to such serious objection. In the past when specialization of interest was less and individual investigators could cover a substantial span of the range from the specialized to the unspecialized, this consideration may have been of no great moment. Now the position is otherwise, for the misconceptions inherent in the traditional classification themselves subscribe to the dangers imposed upon us by specialization. The precondition to all scientific advance is a docile intellectual submission to natural experience. One has only to recollect how chemical knowledge was held up throughout the 18th century by the phlogiston theory, or medical knowledge for an even longer period by the theory of humors, to appreciate this. Nowadays it seems incredible that men could have been so unreceptive to experience that such conceptual obstacles to progress could survive. In general, however, men register the experience they expect, and, if their attention is circumscribed by a mistaken idea, they tend to remain unreceptive to the significance of events outside this for the views they entertain. Yet it is often just such events that break the intellectual impasse which is preventing further development. Anything that renders men less receptive to the significance of any knowledge of natural experience is, therefore, to be most strongly deprecated. A serious objection to the traditional classification of scientific knowledge is that it tends to do just this. By presuming a scale of values in scientific knowledge, it introduces a prejudice into our attitude towards the different categories of natural experience and their significance for the development of scientific knowledge as a whole. Taking these objections in conjunction with the uncertainty that surrounds the decision as to where it is legitimate to apply the traditional hierarchical system to any particular range of subjects, the soundness of this classification is properly open to serious question.[1]

[1] Recently the terms 'Big Science' and 'Little Science' have come into increasing use. These, however, hardly merit the status of a classification.

V

For practical purposes the traditional classification is now proving to be a serious liability. In no case has this been more evident than in the attempts that have been made to use it as a basis for the defence of the essential freedoms of scientific enquiry. It would be worth while, therefore, to look at this example of the difficulties to which it has given rise.

A necessary requirement of all creative activity is that it should have adequate freedom to work out its own development undistracted by considerations which are irrelevant to this purpose. All scholars and all research workers are concerned to ensure this, and their anxiety on this score is not entirely unwarranted by experience. It is understandable, therefore, that scientists, in their loyalty to their ethic, should be suspicious of anything that savours of direction or importunity. But no claim for special privilege, however justifiable, can be sustained without convincing others, and, now that scientists have become almost completely dependent upon society for their material support, this has become of increasing concern to them. To provide a convincing justification for those subjects that can be seen, even by an inexpert audience, to have a purpose that bears directly on some public need, presents no difficulty. To justify to a like audience subjects whose purpose has no such obvious bearing can, however, be a very different matter and, now that the costs of research in certain 'basic' sciences are rising steeply, the claims that they advance for public support have raised the question in an acute form.

Obviously, if the purpose of developing a particular subject is not evident from the nature of the knowledge it produces, then the case for its justification must be based on some other external and independent consideration. In general, the case for supporting research in a so-called 'basic' subject has been defended on

---

Essentially by Big Science is meant science in which research is very expensive because of the facilities it requires; by Little Science, science in which research is relatively inexpensive. Intellectually these terms have no significance but, because of the inevitable connotation of importance in the words 'big' and 'little', they can mislead.

one or more of three grounds.[1] The first of these, whilst admitting that it is difficult to demonstrate any direct correlation between current basic research and national benefit, points to its unexpected benefits in the past and urges what is, in effect, an act of faith in respect of the future. This line of defence is clearly legitimate, although it is unlikely in practice wholly to assuage scepticism if this exists. The second line bases itself on the value of the basic study for educational purposes; the third on its contribution to general culture. Both these defences, however, suffer from the weakness that their appeal is to considerations that are incidental rather than central to the essential purpose of scientific research, namely to develop our understanding of natural phenomena. It is not perhaps surprising, therefore, that anxiety continues to be felt.

Yet there is a case that could be wholly adequate in itself. If a subject can be shown to be an integral part of a continuum, or province of knowledge, which at its periphery engages directly with problems of evident social concern, then it is in no need of other justification. In other words, it is the whole province of knowledge that provides the basis for justification, and subjects derive their individual justification by reason of their participation within this. Considered in isolation, molecular biology might seem (and indeed at one time did seem) to have scant claim for support from funds for medical research. Viewed however in the whole context of biomedical knowledge, it is at once evident that it is an indispensable constituent of understanding. Again the support of research in theoretical chemistry or physics (even including their necessarily large measure of self-orientation) can be substantially justified if considered in the context of the many provinces of natural knowledge to which they make essential contributions. It is because such subjects have been presented divorced from their intellectual contexts, and the attempts made to represent them as existing in their own right, that the present difficulties have largely arisen. But to think in terms of context

[1] *Basic Science and National Goals*: Report to the U.S. House of Representatives Committee on Science and Aeronautics by the National Academy of Sciences. Washington, 1965.

does not perhaps come as naturally to us today as it might have in the past. Owing to the increasing specialization of interest that is being forced upon us by the growth of knowledge, our perspectives are tending to narrow down to the level at which we are working. Against this the traditional classification of scientific knowledge provides little corrective. On the contrary, by obscuring the essential relations between specialized knowledge and unspecialized, it has tended to isolate the latter from its context and thereby not only detracted from its larger intellectual significance but created difficulties for its justification.

We may feel therefore that the traditional classification of scientific knowledge has grave defects. Why then has it such a popular appeal? I believe that there are three reasons. First, it is consistent with our deeply engrained habit of thinking of the structure of knowledge in terms of classical philosophy and on the analogy of a tree. Second, by distinguishing basic research it gives reassuring prominence to subjects in which there is the greatest fear of being overlooked. Third, it has an undoubted value as an arrangement of knowledge for teaching purposes when, if the student is not to get lost in detail, it is essential to devise an orderly progression of instruction from the unspecialized to specialized knowledge.

VI

But if the traditional classification is so faulty, what can we put in its place?

It will probably be accepted without argument that any classification, if it is not to predetermine our conclusions, must be based upon objective and verifiable considerations. To this end, the terms that it uses should relate to facts and not to our judgments on the significance of these. The fundamental defect of the traditional classification is that it fails in exactly this respect. Unless we first make value judgments in regard to the component subjects of knowledge, it cannot be put to use. Inevitably, therefore, when we use this system as a basis for considering the structure and development of scientific knowledge, we are likely only to rediscover the assumptions that were implicit in the categorization we gave to the various subjects before we started. But is it possible to escape from this dilemma? In my opinion it is. To do so, however, will involve our abandoning the traditional attempt to deal with knowledge in the abstract in favour of dealing with it in relation to actual natural phenomena.

In the second chapter of this essay four examples from the biomedical province were given of the sequences of necessary knowledge that were required for their comprehensive understanding. Each sequence ranged from the specialized extreme of knowledge gained from clinical situations to the relatively unspecialized extreme of knowledge gained at the molecular level. None of the knowledges forming this sequence was dispensable. All were required for the over-all understanding. Yet without the particular experience of natural phenomena with which each dealt, none could have been obtained. Consider the range of experiences relevant to our understanding of the significance of Vitamin B_1 for living organisms. At one extreme we have the specialized clinical problem of the disease beri-beri and the specialized epidemiological situation that established the association of this with defective diet. Next we have the pathology of the human disease and the production in animals by means of the defective diet of its experimental counterpart. After this stage the fields of experience could be opened to biochemistry and chemistry; isolation of the missing dietary factor, its identification and synthesis and the elucidation of its role in normal intracellular chemical reactions followed. Again, take the range of experiences that have led to our understanding of the control of bodily growth by the factor secreted by the pituitary gland. This starts with highly specialized knowledge of the clinical situations of gigantism and dwarfism and progresses similarly through less specialized knowledge of the pathological, biochemical and chemical problems that were successively opened up for investigation.

I have chosen these examples at random from the province of knowledge with which I am most familiar, and doubtless others will be able to identify similar examples in the provinces which are their particular concern. But I hope that I have said sufficient to establish my point. Essentially the sequences or provinces of knowledge that we have been considering are the intellectual counterparts of sequences of progressively less specialized problems which are successively uncovered in the investigation of natural phenomena. It is therefore the degree of specialization of the problem that determines the degree of specialization of the knowledge relating to it. And that—the degree of specialization of the problem—is something external to estimations of intellectual value and, as such, something that can be verified objectively.

I propose, therefore, to discard the traditional classification of scientific knowledge with its inevitable presumptions of value, and to continue to use the terms 'specialized' and 'unspecialized'. By these I shall mean no more nor no less than knowledge related to specialized and unspecialized problems respectively. Used in this modest sense, these terms appear free from any implicit commitment to judgment of relative value such as would introduce preconceptions when considering the place of different subjects in the over-all structure of scientific knowledge or their significance for its future development.

VII

At the outset of this essay I stated my belief that only in so far as policy and organization conform to a realistic concept of that which they seek to support will they be effective. Without this they are condemned to move from one expedient to another. In regard to the support of scientific knowledge the reality with which they have to conform is that of the structure of such knowledge and the way this develops. In the preceding chapters I have sought to examine how scientific knowledge progresses and in the present one how it is constructed. This has led me to reject the classical doctrine which sees scientific knowledge as essentially a single entity, science, and the corollary of this that there is a fundamental kind of knowledge on which development depends for its inspiration. This view, in my opinion, derives largely, not from an objective view of the situation, but as a legacy from philosophies that originated in the pre-scientific age

THE STRUCTURE OF SCIENTIFIC KNOWLEDGE

and which, although we may have relinquished their explicit purpose, still remain deeply engrained in our habits of thought. In opposition to this, the classical view, I have put forward a different concept of the structure and categorization of scientific knowledge. In agreement with Karl Pearson, I believe that the only unity of such knowledge, as in logic, lies in its methodology and that scientific knowledge as such is as varied, and as incommensurable, as the kinds of natural experience to which it is related. Further, I cannot accept the view that we can distinguish a fundamental category of scientific knowledge upon which understanding of natural phenomena essentially depends. Instead I believe that, in relation to the different kinds of natural experience, scientific knowledge from the most specialized to the most unspecialized forms along a single continuum and that it is within and through this that its development occurs.

But it is the traditional theory of the structure and organization of scientific knowledge that is tacitly informing current attempts at scientific planning. In the past when knowledge was in its early stages and predominantly concerned with the discovery and analysis of data, this was of relatively little moment. Now, however, when knowledge has advanced to the stage that policy for its further development is beginning to be feasible, our ideas on how scientific knowledge develops are becoming of decisive importance. If, as I believe, the traditional theory is at best a half-truth, and we proceed to organize on the basis of this, then we must anticipate continued disappointment. If, on the other hand, we can provide ourselves with a more correct theory of the structure and development of natural knowledge we can hope for success. In my view, the concept of provinces of natural knowledge that are specialized and separate at their external frontier and increasingly unspecialized and convergent as they approach their central, corresponds more closely to the realities of scientific experience than the traditional concept within which we have been brought up. I propose, therefore, in the light of this concept now to examine the various instruments at our disposal for the advancement of such knowledge and then to consider the question of policy for its further development.

The Organization of Scientific Development

6

THE EVOLUTION OF ORGANIZED RESEARCH

I

THE ORIGINS OF ORGANIZED RESEARCH are to be found in the inherent efforts of man to master his natural environment. It was not, however, until the adoption of agriculture and the consequent institution of settled communities that the first evidence of systematization in his efforts began to appear. Before this, and in respect of many of his occupations for centuries thereafter, man's progress depended on the slow accretion of practical experience and its embodiment in traditional lore. This was the basis of the practical arts and it remained unchanged until the discovery of new sources of energy made deeper understanding essential. But agriculture, and its requirement for communities settled permanently on the land, brought with it from the start the need to supplement experience by understanding. The dependence of crops on the seasons made it necessary to devise an annual calendar and so pointed men's attention to the stars and laid the basis for astronomy. The need to survey the land led to the development of geometry. Barter made weights and measures necessary and enumeration led to the study of numbers. The building of dwellings, stores and temples laid the basis of mechanics and promoted the further development of geometry and primitive mathematics. In each case the incentive to knowledge arose from the emerging requirements of a settled existence.

The new kind of knowledge that thus developed differed in several important respects from the traditional lore of the practical arts. First, it was more than a simple elaboration of practical day-to-day activity. Progressively it became less concerned with the specialized situation of the moment and more with less specialized knowledge relating to all situations of this particular type. It thus began to separate from the practical activities to which it was relevant, and special occupations associated with the idea of learning and its development began to emerge. Secondly, directed as it was to generalization, it was from its beginnings dependent on records. As a result it not only stimulated the development of literacy but became increasingly the prerogative of the literate. And thirdly, were the consequences of its concern with natural phenomena. Primitive societies have always seen in natural phenomena the manifestations of supernatural agencies. It was, therefore, inevitable that the early development of natural knowledge should become associated with religion.

By the Egyptian period these developments had become established. The idea of systematic knowledge was identified with authority and prestige. Learning was becoming progressively concerned with abstraction and was drawing increasingly away from concern with the practical and technical arts. In the succeeding Grecian age, although the supernatural element declined in importance, the withdrawal from practical affairs continued so that it became almost a point of honour among the educated elite to disavow technical interest. Throughout the Roman and mediaeval periods this outlook persisted so that by the end attention was concentrated largely on the working over of ideas and, among the learned, actual investigation of natural phenomena was practically in abeyance.

To suggest this as a reproach is to be unrealistic. Agriculture still set the material limits within which civilization could develop and, in consequence, the problems of the natural environment that exercised men's minds. Until almost the end of the first millennium and a half, these were no different in kind from those that had faced men throughout the ages and as such they provided no incentive to go beyond the elaboration of established practice. Then, with the invention of gunpowder and with the undertaking of oceanic voyages the whole situation altered. New dimensions were introduced into human experience and with it fields of interest that had previously been confined to the relevant practical arts began increasingly to attract intellectual attention and to provide new incentives to the development of systematic knowledge. As a result, the traditional gap between learning and the material concerns of practical life started to narrow.

In the course of development previous to this stage, the position of medicine as the only biological subject with pretensions to learning merits some attention. To some extent this, in the catholic sense in which medicine was then understood, provided a link between the practical and the abstract. Embedded in the church on the one hand and engaged with natural experience on the other, and comprehending as it did primitive biology and chemistry, it served as a bridge between two worlds. As such it was the main way open to educated men who wished to become acquainted with natural philosophy in any of its branches. It should be no matter for surprise, therefore, that men like Copernicus[1] and Galileo[2] read medicine at the university; that a President of the Royal College of Physicians[3] was the father of the science of magnetism and electricity, or that out of the nineteen founder members of the Royal Society who could be regarded as actual scientists, fourteen were doctors of medicine.[4]

This then is the general background against which we have to view the evolution of the organized development of scientific knowledge in modern times. The period in question starts some eight hundred years ago with the emergence of universities. It is marked in its middle stages by the appearance of scientific societies and, in its modern period, by the creation of national organizations for the promotion of scientific research. The evolution of each of these and the circumstances that brought them successively into being, now fall to be considered.

---

[1] Herman Kesten: *Copernicus and his World*, p. 110. Martin Secker and Warburg, London, 1945.

[2] Phillip Lenard: *Great Men of Science*, p. 26. English translation by H. S. Hatfield. G. Bell & Co., London, 1933.

[3] *William Gilbert*. Munk's Roll of the Royal College of Physicians, 1878, Vol. 1, p. 77.

[4] Henry Lyons: *The Royal Society, 1660–1940*, p. 22. Cambridge University Press, 1944.

II

The universities in the sense that we now know them arose originally from professional schools.[1] The first organized centre of higher education that we know of in Europe was the medical school in Salerno which was well established by the 11th century. This was followed by the predominantly legal school at Bologna. Throughout Europe, however, schools had developed in association with religious institutions. To regard such cathedral schools as catering only for those bent on a purely ecclesiastical career would be mistaken. In those times the church was the main source of men with the training for administering the rudimentary machinery for social organization that then existed and the sole source of those with any higher learning. Further, knowledge was still so limited that specialization was as yet hardly more than foreshadowed and alternation or combination of ecclesiastical and lay functions was still possible. From the educational activities of the church there came, therefore, not only the clergy and administrators but, in the earlier times, most of the lawyers and physicians.

By the 12th century the inadequacies of this system were becoming apparent and the idea of the *studium generale* where all subjects could be studied at the higher levels and to a generally acceptable standard of proficiency began to emerge and the university in its modern sense to take shape. Important as this change was, it was no drastic innovation. Owing to the limitation of knowledge, and specialization being as yet little more than an emphasis in interest, the idea was already latent in the main cathedral schools. Its significance lay rather in the tacit recognition of education outside the immediate aegis of the church and the facilitation that this gave to the pursuit of knowledge for secular ends. Despite this it was some considerable time before the lay professions were separated from the church. Not until the early 16th century did Linacre effect this for medicine in this country[2] and it was only later that the lay administrator, as

[1] H. Rashdall: *The Universities of the Middle Ages*. Clarendon Press, Oxford. New edition, 1931, Vol. 1.
[2] H. P. Himsworth: *Lancet*, 1955, ii, 217.

distinct from the ecclesiastic, became the pattern in government.

It was from such beginnings that our major instruments for higher education, the universities, have evolved, and inevitably this has considerably influenced the way they have developed. In particular, it has largely determined our concept of the university as an institution and the fields of interest with which it is engaged.

To the fact that the universities developed on the basis of the three professions of the church, the law and medicine, and not from the practical or commercial activities of society, we can attribute the characteristics of their ethos. Despite the difference in the subject matter with which these professions deal, each has developed traditions and ethical codes that are essentially similar. The reasons for this are not far to seek. All profess a recondite body of knowledge. All deal with personal matters of intimate and vital importance to those concerned. All depend ultimately on trust in the knowledge and integrity of their individual members. When, therefore, the university emerged, it inherited the professional traditions of autonomy, responsibility for the development of its particular knowledge and an exacting ethic of individual integrity. It is from this basis that the concept of academic freedom has developed with its corollaries of institutional and individual autonomy and its tradition of research.

The professional origin of the universities has also been responsible for the balance of interest within them. Concerned as they primarily were with the preparation of men for the church, the law, medicine and administration, it was inevitable that their staffs should be predominantly composed of men whose knowledge was relevant to this purpose. In consequence, those many fields of interest that were related to practical affairs and the world of commerce lay largely outside their consideration. At the outset, therefore, they tended to perpetuate the distinction that had grown up between learning and practical affairs. For the first half of their existence this was of little importance. When however towards the end of the 15th century the new discoveries and inventions of practical life that stimulated the development of the physical and material sciences began, the universities as such remained largely on the periphery. The result was that until the early part of the 19th century the major development of the new sciences took place largely outside them.[1] Only in those branches of scientific study that had, before the inception of the universities, established their position in learning, such as mathematics and astronomy, was the condition otherwise. Although, therefore, most of the great scientists of the period had attended universities as part of their education, the development of their scientific knowledge was generally subsequent and independent of this, and even those who remained in the university had to look outside for others with congenial interests. It thus came about that in the first great period of the scientific revolution the universities remained generally aloof. In consequence, men turned to other devices.

The first attempt was to found another kind of college. In this country, in the latter part of the 16th century, a London merchant, Sir Thomas Gresham, attempted this very thing. His scheme was clearly no mere incidental of civic pride but a conscious and deliberate attempt to meet the changing needs as he saw them. There were to be no students but instruction was to be given by the seven professorial departments to the interested lay public and, typically, at least half of this was to be given in English rather than Latin. The attempt was presumably premature, for, nowadays, Gresham College is mainly remembered as having been the meeting place of those who later founded the Royal Society. Two hundred years later, however, Rumford's not dissimilar scheme for the Royal Institution provided the opportunity for the genius of Davy and Faraday.[2]

[1] Eric Ashby: *Technology and the Academics; an Essay on Universities and the Scientific Revolution*. Macmillan, 1966.
Eric Ashby: *Universities, British, Indian, African*, Chapter 1. Weidenfeld and Nicolson, London, 1966.

[2] Although Gresham College may have been premature, it seems possible that it provided the idea for the salaried appointment of a lecturer by the Royal College of Physicians. This lectureship was held by William Harvey from 1615 to 1656, and whilst holding it he did the work that made him and this country famous in medicine. As the scheme for Gresham College could not be implemented until after the death of Lady Gresham, the medical

But the effective device for the needs of the time proved to be along other lines. This was the scientific society where men of like interests could meet, exchange knowledge and promote experiments, and publish their records.

III

The 17th century saw the foundation of scientific societies in Italy, France and this country, and we might perhaps be pardoned for believing that the most influential of these has been the Royal Society of London. These early societies took all natural knowledge for their province. Further, they were inspired not only by curiosity but also, perhaps under the influence of Bacon, by a genuine belief that their experimental philosophy should promote the material welfare of men. Hooke, for example, clearly defined the business of the Royal Society in his preamble to the Statutes of 1663 as: 'To improve the knowledge of all natural things, and all useful Arts, Manufactures, Mechanick practices, Engynes and Inventions by Experiment.'[1]

Since these early days the device has been widely followed, so that now every developed country has an academy or its equivalent as its senior scientific body. With the passage of time, however, and the enormous growth of scientific knowledge, the possibility of any single body of men giving adequate consideration in detail to all the scientific subjects that were developing disappeared. Scientific societies with interest restricted to particular subjects then began to appear, so that today societies for chemistry, physics, biochemistry, physiology, geology, to mention examples of only the major of these, have become accepted features of scientific organization.

The scientific societies are even more indispensable today than they were at their foundation. By providing a forum for informed criticism they sustain the standards of scientific work. By bringing together a diversity of relevant experience they facilitate the

---

lectureship antedated it and became thereby this country's first research post. H. P. Himsworth: *British Medical Journal*, 1962, ii, p. 1537.
[1] H. Lyons: *The Royal Society, 1660–1940*, p. 41.

generalization of knowledge in regard to their particular subject. It is to such societies that we mainly owe the promotion of theoretical understanding in those subjects, like chemistry and physics, that stretch like strata across sequences and provinces of knowledge at the more unspecialized levels.

IV

By the early part of the 19th century the tide of scientific development was flowing strongly. The scientific societies were firmly established and the interest that they fostered was spreading in widening circles. The industrial revolution permeated the whole outlook of the commercial community and gave incentive and confidence to inventive talent. In this country the Royal Institution and the Royal Observatory and the personal researches of private individuals were systematically delving to deeper levels. The development of empirical philosophy and the growth of utilitarianism were influencing the whole attitude towards new knowledge.

The invasion of the seats of higher learning by the new scientific knowledge could not be long delayed, and with the establishment of the University of London (later University College) at the end of the first quarter of the century, this found effective expression in this country.

The history of the organization of scientific knowledge during the next hundred years is that of the rapid development of these lines. With the emergence of new subjects, further and more specialized scientific societies were created. New centres of higher education, in which the new scientific knowledge was strongly represented, were set up. In the older universities, existing departments were adjusted and new departments progressively created to accommodate the new trends. The whole field of natural knowledge was expanding at an unprecedented rate from its specialized periphery to its unspecialized centre and by its very success it was raising new problems both for its further development and for society.

By the early years of this century it had become evident to

THE EVOLUTION OF ORGANIZED RESEARCH

thinking men that scientific knowledge was transforming our civilization and that no country could hope to keep in the forefront unless it made deliberate provision for research. Further, with the accumulation of success and the increasing certainty of technique the attitude of society to research was altering. No longer was it being regarded as a recondite activity of a few rare individuals that might, occasionally and somewhat unpredictably, yield gratuitous benefits. Rather the belief was gaining ground that given proper organization and support scientific research could, and should, be undertaken with considerable assurance of success. The situation was changing also for research. Its costs were rising rapidly and were already threatening to exceed the resources of endowment or private benefaction. More assured and systematic support was becoming a necessity. Further, the research effort was divided and scattered and the need for co-operation was, with the emergence of multidisciplinary subjects, being increasingly felt. A new situation was clearly developing, whether this was regarded from the point of view of the public or from that of scientific interest. To meet this, some additional device was required, and it was for this purpose that central research organizations, supported from public funds, were brought into being and charged, on the one hand, with promoting research over the whole range of knowledge relevant to their field, and on the other with providing society, as expressed by government, with the scientific knowledge and judgments required.[1] That such nation-wide research organizations have been created in all modern countries and that development has followed a very similar pattern in each indicates that, like the universities in the late middle ages and the scientific academies over three centuries ago, they in their turn are a response to a further stage in the progressive development of knowledge.

The present position is, therefore, that in the course of the evolution of scientific knowledge, four instruments for its advancement have come into being: the professions and industries, the universities, the scientific societies and the central research

[1] *Report of the Machinery of Government Committee*, 1918. Cd up to 1919 9230. H.M.S.O., London.

organizations. Each of these has been brought into existence by needs that successively emerged as a consequence of the development of scientific knowledge itself. In no sense is any one a substitute or replacement for any other. All have their particular contribution to make and all are necessary to meet the needs of the stage that scientific development has now reached. The role of these different instruments in the modern situation, and even more so their role in that which is opening before us, must therefore concern us closely.

* 7 *

THE UNIVERSITIES AND RESEARCH

I

DURING THE NINETEENTH CENTURY the universities of the western world moved increasingly to the forefront of scientific advance so that by its end they had become beyond question the centres that were providing its leadership. It was not only that under their aegis the new physics and chemistry were developed with a speed and brilliance that carried understanding to undreamed-of heights, or that they provided an increasingly sympathetic milieu in which the new confidence and discipline could spread to other studies. It was also that, because of their traditional responsibility for higher education, they were able to ensure that, in parallel with the developing scientific knowledge, increasing numbers of men trained in the new disciplines should come forward. There can be no question, therefore, that over the last century and a half the university as an institution has proved itself in relation to the scientific revolution.

In the atmosphere of acceptance of the present day, it is easy to take for granted that events could have followed no other course. Yet, if one looks at the state of the universities in the 17th and 18th centuries and their relative detachment from the intellectual developments around them, it might be felt that this was by no means a foregone conclusion and that, in response to the changing situation, some alternative type of organization might well have come into being. But this did not happen. Instead, the universities adapted themselves to the new conditions and, without losing their essential character, moved forward to take the lead. It would seem justifiable, therefore, to infer that embedded in the concept of a university there must be certain principles of continuing value for the promotion of research and that, whatever the circumstances, a country's institutions of higher education will have an indispensable role to play in its development.

But no institution can exist for any length of time without accumulating traditions of varying validity. Today, when circumstances in regard to universities are changing rapidly and many traditional assumptions are under challenge, it is more than ever important that we should be clear as to the features that enable them to make their particular contribution to the over-all advancement of knowledge. I propose, therefore, to look at the universities from the point of view of their role as an instrument of scientific research, with the object of attempting to clarify first the particular nature of the contribution to be expected from them, secondly the conditions that are necessary to enable them to make this, and finally the new circumstances in which they find themselves and the possible effects of these upon their future role.

II

The essential purpose of the university is higher education. It is this that distinguishes it from all other institutions and constitutes the basis of its claims to special consideration. Although, therefore, in the discharge of this responsibility universities have acquired further interests, it is to the requirement of higher education that we must look ultimately for their justification and characteristics.

Higher education can be distinguished from the other forms of education that precede it by its concern being rather at the frontiers than the hinterlands of knowledge. It requires, therefore, that those engaged in providing this should be constantly and critically aware of the new accretions to knowledge as these occur, and that, to this end, they themselves should be actively seeking, by scholarship or research, to contribute to the body of knowledge with which they are concerned. The question as to what kind of knowledge is relevant to this purpose is, therefore, an important one.

The requirements of higher education necessarily determine both the representation and the balance between subjects within the university. In education it is a well-attested rule to proceed from the less specialized knowledge to the more specialized. Thus an intending medical student will enter on his course at the unspecialized levels of biology, chemistry and physics before proceeding to human anatomy, physiology, biochemistry and biophysics, and only after he has traversed these to the extent required will he enter on the study of the abnormal. Again, an engineer, whatever his ultimate specialty, will start through the physical sciences and mathematics. Thus of necessity the unspecialized subjects have to be strongly represented in the university, not only in the interest of those who will need to make use of them in their further studies, but also for those who intend to devote themselves to knowledge at the unspecialized levels themselves. In consequence, as recruitment to the staffs of universities must reflect the educational requirements, the balance of research effort within any province of natural knowledge tends to be towards the unspecialized. We may expect, therefore, that the contribution of universities to scientific research in general will be proportionately more at the unspecialized than at the specialized levels of knowledge.

But the requirements of education have also a bearing on the qualitative as well as the quantitative aspects of research in universities. In that it seeks to prepare men to deal with the diversities of future experience, higher education is necessarily more concerned in any subject with the general rather than the particular. In consequence we should expect what we in fact find, that research in universities tends to be more concerned with developing the conceptual framework of a particular subject than with its expression in particular situations or in relation to the experience from which it derives. This applies at any level of knowledge from the specialized to the unspecialized. Thus pathology in the universities is as much concerned with the general principles it can extract from its experience as biochemistry and molecular biology are with theirs, and even academic departments as specialized as those of chemical engineering are orientated more to developing the unity underlying the diversity of their operations than to following out the ramifications of these. To say this is to say no more than is meant when it is said that the particular concern of universities is with basic knowledge and basic research.[1]

The influence of higher education on the research associated with it would seem, therefore, to be twofold. Quantitatively it would tend to incline the balance of effort towards the more unspecialized. Qualitatively it would tend to focus attention on the conceptual or theoretical aspects of any particular subject. Looked at against the background of the over-all requirements of scientific knowledge, it would appear, therefore, that the contribution to be expected from the universities will tend to be weighted towards the unspecialized centre of scientific knowledge and towards the development of unification at the subject level. To these particular ends the organization of a university is well adapted.

III

The traditional concept of a university as an organization is that of a self-governing community of scholars each of whom has certain claims to autonomy in professional matters. The justification for this concept and the doctrine of academic freedom rests on the individual nature of responsibility in teaching and the development of knowledge to this end. In respect of research, therefore, each independent member of the academic staff has the expectation not only of doing research but also of determining within the general ambit of his subject what research he will do. The feature of research in universities is, therefore, that it is, in principle, essentially a matter for individual decision and such it remains even when the individual comes increasingly to require support by assistants or even a whole departmental team. This traditional principle has important corollaries.

Subject to the general proviso that the research undertaken

[1] The extent to which academic representation is disposed towards the unspecialized varies, of course, with the sequence of knowledge. In respect of those sequences like medicine or engineering that are represented more or less in their entirety, the situation is naturally less selective than in those like physics where representation is confined to component subjects.

shall contribute to the knowledge that he professes, it is entirely a matter of individual judgment whether a particular line of investigation shall or shall not be undertaken. That proviso apart, the individual academic worker is under no other obligation save those that he voluntarily accepts. This situation has consequences that are both highly rewarding and limiting. By ensuring as it does that there shall be a multiplicity of outlets for initiative and creativity, it provides wide opportunities for the emergence of new ideas and new talent, a safeguard against the persistence of individual error and oversight and, by the very diversity of the interests over the universities as a whole, an insurance against the unforeseen. Further, it ensures for society a wide range of independent opinion which, because of its relative insulation from the obligations of practical responsibilities, can be drawn upon in matters of doubt or controversy. These are indispensable provisions. Collectively they provide a safeguard against the persistence of wishful thinking and a spread of investment against future needs. Academic freedom and university autonomy which make these possible are, therefore, no mere survivals of mediaeval privilege, but rather essential requirements under any conditions.

IV

Although the basic strength of the universities as instruments for scientific research may lie in their being a means to ensure multiple centres of independent initiative, this is not, however, the only reason that has been put forward for their suitability as research centres. In addition, it has been maintained that the universities, *qua* universities, provide an environment that is unrivalled for the development of knowledge and further that the conjunction of teaching and research is little short of essential for this purpose. Both these contentions deserve, therefore, our serious consideration.

The view that research is best placed in an academic environment is based on the experience that highly specialized research institutes set up in isolation do not as a rule continue to flourish. As research nowadays is so commonly multidisciplinary, this is not surprising. But one has only to look at the larger multidisciplinary research institutions like the National Institute for Medical Research, the Rothamsted Experimental Station and the National Physical Laboratory in this country and the National Institutes of Health in the United States to see that the essential element is not the academic nature of the institution but the fact that it is a multidisciplinary community of adequate size and diversity. It would appear, therefore, that it is to the universities providing communities of this nature, and not because they are universities, that their intellectual suitability as an environment for research can be attributed. But in this respect they will not be unique.

The statement that research and higher education are indissolubly linked can be looked at from two different points of view. One is that research is a necessary adjunct to higher education. The other is that higher education is a necessary adjunct to research.

On the basis of the considerations in the previous section, the statement that research is a necessary adjunct to higher education can, in principle, be accepted without any question, although the detailed implications of this will require further discussion. No similar support can, however, be adduced for the second proposition. As we have seen, the foundations of modern scientific knowledge were laid largely outside the universities and by men who were not concerned with education. Further, over the last two decades, nearly half the Nobel Prizes awarded in biomedical research to workers in this country have been won by men who were not members of academic staffs. In the face of this evidence the contention that engagement in teaching is essential to the carrying out of high-quality research is clearly untenable. That, of course, must not be taken to imply that teaching is in any way inimical to research or incompatible with its pursuit. On the contrary, many research workers find teaching to be stimulating and a salutary counteraction to the risks of excessive specialization. In respect of university staff, teaching raises no problem, for by accepting an academic appointment they have *ipso facto* incurred teaching obligations. In regard to research workers not on academic staffs there is, of course, no such obligation, but, in practice, most would like to teach at least at the postgraduate level.

In this and the preceding section I have been concerned to establish the unchallengeable element in the entitlement of universities to do, and therefore to be supported in doing, research. I am aware that many different reasons have been advanced in support of this contention but, on examination, only two of these seem to have unquestionable validity. These are first that research is a necessary adjunct to higher education, and second that the universities, because of their obligation to the disinterested pursuit of knowledge, are a major instrument for the development of intellectual initiative. These apart, members of the academic staffs of universities, *as academics*, have no special title to consideration in respect of research. That is not to say that they may not have other titles. Thus a physician in the medical faculty of the universities is, like any other physician, under an obligation to do everything he can to improve his professional knowledge and thereby his practical ability. But that is a professional not an academic obligation, and if used as the basis of a claim for support in research, falls to be judged on medical considerations, not academic. The claim of universities, as universities, to receive support to enable them to promote research rests, therefore, on those characteristics that arise from their being institutes of higher education and any request for support beyond this must rest not on entitlement, but on other and external considerations.

V

Research in university departments is supported in two ways. One is from the general funds of the university. The other is from external sources such as the relevant publicly financed central research organization, and, to a lesser extent, the individual private research foundations or industry. In this country the University Grants Committee, in making its block allocation of public funds to universities, takes into account not only their requirements for teaching but also those for the promotion of research. Thus, in accordance with the principle that higher education must be associated with research, the universities are assured of funds, at their own disposal, to maintain a basic level of research activity within their departments. If, however, the research in any department, or by any particular worker within it, shows signs of needing to develop on a scale beyond the resources available locally, the understanding is that the worker concerned will apply to an appropriate external source for the extra support required. Of these the central research organizations, by the size of their contributions, now so outweigh support from other sources as to have become the decisive factor in this respect. Although therefore contributions from private sources for research in universities are still a significant item, the problem today is essentially that of support from public funds and, in this country, the dual system by which these are channelled through the University Grants Committee on the one hand and the individual central research organizations on the other.

In principle this dual arrangement provides for the essential needs of the situation. By ensuring for individuals a basic level of support for their investigations, it enables them to exercise their own judgment in testing out their new ideas, and thereby ensures to the country a multiplicity of centres of independent initiative. By making available external sources of support to which they can have recourse, it provides a safeguard against the danger of promising research being curtailed by local financial stringencies. It would be idle, however, to expect that any arrangement, even when it is fully accepted in principle, would necessarily meet with universal approval in practice. The volume of research, its costs and the scale of its individual operations are rising exponentially. In aggregate they are already approaching formidable totals and it is clearly unrealistic to assume that the present trend can continue indefinitely. Yet to any individual research worker, his particular interests (or if he is more senior, those of his particular institution) are naturally of first importance. Although, therefore, the system of financing research in universities through two separate channels is generally approved in principle, the scale upon which it operates will inevitably be criticized and considerable differences of opinion are to be expected on the question of the proper proportions between support from internal and external sources. These are both important points.

Obviously the funds available for the support of research in universities through both channels together will depend on the financial resources of a country and the amount of these that it has been decided shall be devoted to this purpose. The first of these factors is clearly outside our consideration; the other can better be discussed in the context of the arrangements for the overall support of scientific research. The second question, however—that of the desirable proportion of support from the two sources—requires consideration now. It can most easily be approached by asking: 'What is meant by a basic level of support?'

If the characteristic contribution of the universities to the over-all resources for research in a country is the provision of a multiplicity of centres of intellectual initiative, then evidently these must have, at their own disposals, funds sufficient to make the individuals on their staff independent of external approval in the initiation of their research. That is not to say, however, that such funds should be large enough to provide for the extended development of all the lines initiated. Not all research proposals are equally meritorious or important. Yet, human nature being what it is, it is not easy for a local community to question a colleague's judgment, especially when the basic traditions of such a community include that of the intellectual autonomy of the individual members. Being realistic, we must draw a distinction between the initiation of a line of research and its further development. Whilst, therefore, it is only enlightened self-interest for a country to ensure that its universities have sufficient money at their unfettered disposal to allow new lines of research to be initiated, it is neither in its own interest, nor in that of the advancement of scientific knowledge, that the continued development of lines of research should be isolated from effective expert scrutiny by making them entirely independent financially. The corollary to the claim of local freedom of initiative is, therefore, that the stages of development after the initiation should be subject to an external assessment. Nowadays, when knowledge has become so complex, the scientific assessment of any major research development is unlikely to be adequate at less than the national level. In present circumstances, independent central research organizations where a man's achievements are assessed by those he accepts as his scientific peers and to which he can look for the support required in the further development of his work, have now become indispensable. Without them, it is unlikely under modern conditions that academic autonomy in respect even of the initiation of research could for long continue to exist.

To attempt to lay down the proportions of the support for research in the universities that should come from them on the one hand, and national research organizations on the other, in precise terms, would be a mistake. The proportions that were thought appropriate today might not necessarily be so tomorrow. In general, however, research at its initiation is relatively inexpensive. It is its expansion that is becoming increasingly costly. It is probable, therefore, that the proportion of support coming from external sources will increase with time. Nevertheless, if the universities are to continue to make their contribution, rising costs notwithstanding, it is essential that the resources at their disposal do not fall below the level required to maintain their ability to initiate.[1]

Against the background of the considerations in this and the preceding sections, we can now proceed to look at the present situation.

VI

The new factors now facing scientific research in the universities are: the increasing sophistication and consequent increased cost

[1] In this country a scheme has been devised whereby research previously supported by research councils can, when it has become an established and continuing part of the university structure, be taken over by universities. The incentive to doing this was that the research councils had come to feel that the amount of money for research at the disposal of the universities was falling below the level required to maintain the necessary degree of local initiative.

of the facilities that scientific research now requires; the demands that it now imposes on the time and energy of the individual who wishes to engage in it; the increasing specialization that is being forced upon us by the continuing growth of knowledge. These factors are new only in a quantitative sense. Nevertheless, in relation to organizations, there comes a time when the quantitative increase of their functions brings them under such strain that it becomes questionable whether mere expansion or multiplication of existing arrangements is any longer a sufficient answer to their problems. If we look at the traditional assumptions on which our system of universities has been constructed, and then at the new factors bearing on this, we may well ask if such a point has not now been reached.

Traditionally, all universities are equal in the sense that all men are equal in front of the law. Each aims to provide as comprehensive a coverage as possible of existing knowledge. In pursuit of this, each has the right to obtain, to the best of its abilities, the support for its realization. Conceptually, as we have said, the university is a self-governing community of scholars. Its component units are the various departments of knowledge, each of which is organized on a hierarchical basis epitomized by the traditional, and still generally prevalent, uni-professorial department. When the techniques of scientific research were relatively unsophisticated and cheap, when its demands on the individual were such that it could be undertaken incidentally to his educational duties, and when knowledge was not developed beyond the stage at which one man could oversee the research efforts of the whole department, this scheme was entirely adequate. Now, however, these conditions have all changed, and in consequence the situation has altered.

The impact of the increased technical sophistication and rising costs was first felt with its full force in the subject of nuclear physics. It was clearly out of the question to consider providing a department comprehensively equipped with the necessary massive engineering projects and the highly trained personnel to run these, at every university in the country. Indeed it has proved impossible (other than in the U.S.A. and the U.S.S.R.) to provide the largest machines save on an international basis. Now computers are following the same course, and there are already indications that biologists, in respect of some of the facilities they desire (such as 'mega-mouse' colonies for genetic purposes), are beginning to think on a comparable scale. But these are only extreme examples. In every field of research the cost and sophistication of facilities are rising rapidly, and even in single fields the aggregate over the universities as a whole may well achieve levels at which it has a significant influence on the development of policy both directly for the field itself and indirectly for others.

Were the considerations only financial, they would still be serious. But there are other implications. Complex facilities require highly trained and specialized personnel both for their operation and maintenance. Not only does this draw heavily on scarce manpower but it also imposes on the investigator himself an exacting managerial role which he must discharge if he is even to approach his primary purpose.

This is the present trend of events, and although it may not yet have significantly affected all scientific subjects, it seems improbable that ultimately any will be totally exempt. It is no wonder, therefore, that there has been talk of selective concentration, either on the basis of a particular university concentrating on the special development of at most some very few subjects, or, more radically, of concentrating the more sophisticated work in the generality of scientific subjects at some few centres in any one country.

The other two factors both centre on problems that today confront the member of a university staff who aims, as practically all do, to undertake serious research. It is a commonplace to say that the amount of knowledge in any major subject is now so large that no man can hope to be an expert on more than one aspect of it. Yet the university department of which he is a member has, of necessity, to give (at least to the undergraduates for whom it is responsible) instruction in all aspects. In the traditional uni-professorial department of the customary size, this means perforce that the individual staff member has to teach on, and keep up-to-date in, aspects of his subject in which, as his research develops, he inevitably comes to have a less vivid interest.

THE UNIVERSITIES AND RESEARCH

In the past, when knowledge was less and research less exacting, this entailed no great burden. Now, however, the situation is often felt to be otherwise. The present tendency to increase the size of university departments so as to offset the teaching load on individual members, and so allow more time for research, is the natural response to this. Further, the traditional uni-professorial department is increasingly being represented as an anachronism, and in its place the case for multi-professorial departments with correspondingly large staffs and facilities for research is being strongly pressed. Yet is it realistic to believe that any country could provide for all scientific departments in each and all of its universities to follow such a policy even if it had not, as we in this country recently have, sanctioned a sudden and great increase in the number of its institutions for higher education? Yet as long as every university department aims in principle (as under our existing tradition it has every right to aim) at reaching the highest level of development, the dilemma will be insoluble.

I would not presume to enter into a discussion of the fields of higher education if it were not that such education and research are inextricably bound together. As it is, anyone who would attempt to consider the problems of supporting research in a country has no option but to do so. These, then, are my personal views in the context of the present situation.

Although it might have been avoidable in the past, a distinction might now be drawn between provision for undergraduate and provision for postgraduate education. In this country undergraduate education aims to take a man up to his first degree in three years. Even allowing for the exceptional degree of specialization in the upper forms of British schools, this is not a long period, and as knowledge grows it will appear to be increasingly shorter. Of necessity a high intensity of assimilation is required from the student and a corresponding intensity of instruction from his teachers. It might of course be argued that this is all the more reason why the best brains in the country should be involved in undergraduate teaching. But intellectual stimulation of those who are capable of receiving it does not necessarily imply continuous exposure, and to insist that our best brains should be assigned to duties that inevitably contain much routine is, I would submit, to take a somewhat doctrinaire attitude. The situation has moved on since the days when a first degree in science brought men to the frontiers of knowledge. Today the frontiers are well beyond this, as shown by the frequency with which it has been felt necessary to introduce further courses of instruction into the requirements for the research doctorate in philosophy (PhD).

The situation is, or should be, different in regard to postgraduate education. I must explain, however, that I do not use the terms 'undergraduate' and 'postgraduate' in relation to education as necessarily synonymous with the existing arrangements to prepare men for undergraduate and postgraduate degrees respectively. The distinction as I see it is that a man is still *in statu pupillari*, and not at the stage to be a postgraduate, if he is still in need of courses of instruction. In my view, if the first degree course is insufficient as a preparation for higher work, then the proper thing to do is to devise a course leading to a Master's degree and to include the necessary supplementary instruction in this. This would allow the removal of courses of instruction from the doctorate of philosophy and thereby permit this to be accepted as evidence of what it purports to be, namely that the holder is more than instructed and has shown noteworthy qualities of originality and intellectual mastery. Surely, if we are to make the best use of the best brains available in higher education, it is at this stage that they can most profitably be concentrated.

Earlier in the chapter it was stressed repeatedly that research was a necessary adjunct to higher education. I myself have no doubt that at the level where formal instruction has been left behind this is unequivocally so. Further, at this level the research interests of the teacher are not only in line with his educational responsibilities but further fortify his ability in respect of these. At the undergraduate level today, the case is more questionable. At this level, some of the best teachers are those who are no longer personally preoccupied with opening new and specialized fields of knowledge, and provided that the undergraduate is from time to time exposed to those who are, the range of his intellectual experience is unlikely to be substantially different from that which, at this stage, he is able to assimilate and may, in some cases, be even better balanced.

The implications of these considerations for the problem of universities considered as centres for scientific research and the training of research workers will by now be evident. It is that we should have the courage to recognize the need for two types of university. One would be predominantly (and I hope eventually exclusively) concerned with postgraduate education and might well be called a graduate university. The other would be predominantly, although not necessarily exclusively, concerned with undergraduate education, including in this category those who were taking Master's degrees. Clearly the number of graduate universities would be considerably fewer than the number of undergraduate. At the graduate university the major facilities and major departments for research could be concentrated. I have myself little belief in the proposition that every university might have one or so departments that it develops on a major scale and that then, over the country as a whole, a full coverage will be achieved. The progress of scientific knowledge is characterized by the springing up of new subjects at the junction of the established ones. Further, it is becoming increasingly multidisciplinary. The case for having no gross imbalance between the scale and development of contiguous departments in a university is consequently a strong one. If, therefore, agreement could be reached to set up, or work towards, graduate universities, all their component departments should be graduate also and their scale and facilities correspondingly major.

VII

At the beginning of this chapter it was pointed out that, during the last century and a half, universities had not only provided outstanding leadership in the development of scientific knowledge but also that, because of their responsibility for higher education, they had been able to ensure that increasing numbers of men trained in the new disciplines continued to come forward. Today, when scientific research has become such an essential feature of our civilization, the training of scientists in sufficient numbers has become even more a matter of national concern. The changing situation in this respect deserves, therefore, continuing attention.

The growth of knowledge has necessarily brought educational problems in its train and with its further growth these will inevitably intensify. Because of this growth, specialization with its consequent narrowing of interest is being increasingly forced upon us. The nearer one approaches the frontiers of knowledge, the more intense does the specialization of interest become. In the preceding section we have already touched upon the increasing difficulty in compressing into the three years of the first degree course the instruction now required to equip a man to embark on genuine postgraduate education, and the increasing need for supplementary courses of instruction at the level of the Master's degree. In general, the need for such supplementation, when it involves the acquisition of a further intellectual technique or unspecialized knowledge ancillary to the student's particular field of interest, is generally appreciated both by educators and intending research workers. The inadequacy to which I now wish to draw attention is, however, of a different kind.

In research, it is a matter of common experience that there are some men who seem intuitively always to take the right path whilst there are others who seem always to fall short of expectation. Of course in many cases there is an evident difference in ability to explain this. In a substantial number of cases, however, this is not so. Searching for an explanation of this, I have been struck by one thing. In general, the wider the man's awareness of the extent of knowledge to which his particular interests are related, the more effective his research. This is not altogether surprising. At the end of any investigation, several possibilities have usually opened up and further progress depends upon choosing the correct one for further research. In this the investigation just completed is little help (for otherwise there would be no problem). The choice turns upon the particular investigator's assessment as to which of his several conjectures is most consistent with other relevant experience. Clearly the wider the man's awareness of relevant experience, the more certainly does he narrow down the possibilities and the more likely is he to choose rightly. It is, therefore, of considerable importance in the training of research workers to ensure, if at all possible, that they should be not only aware of but confident in their grasp of the meaning of findings in fields of experience relevant to their own. The question is, what are those fields?

In discussing the way scientific knowledge developed, it was pointed out that three processes come to bear on this. The first is an elaboration of the ideas and data of the subject itself. The second is the application to its material of knowledge and techniques drawn from the more unspecialized subjects in the particular province of knowledge. And the third is the impact of the problems that came down to the subjects from levels that are more specialized and in closer contact with unabstracted experience. All are indispensable, but it is to the latter that the subject primarily looks for the extension of its field of relevant experience. Take the biomedical field, for example. Diseases are natural experiments. Repeatedly, experience derived from their investigation throws unexpected light on normal function and provides information that could have been obtained in no other way. Further, the comparability of ideas on function derived from experiments on animals with ideas derived from the study of the abnormal is frequently a rigorous test of their validity. The ideal is obviously that a biomedical research worker's knowledge should comprehend the relevant aspects of both the normal and the abnormal. Today, however, to require every worker in the biomedical field to take a medical qualification in addition to any degree in science that he may possess is impracticable, even if it were necessary. But the need to offset the tendency to intellectual isolation inherent in the present situation remains, and with further growth of knowledge and the natural tendency of academic interest towards the unspecialized, it can be expected to increase. To meet this, a sufficiency of intending research workers who are medical graduates will need to be trained in the more unspecialized sciences relevant to their particular interests, and a sufficiency of those who are science graduates to be given some introduction in medicine. Means to realize the former of these requirements already exist and are being used increasingly. The latter need has however as yet received little recognition, and measures to meet it are non-existent. Yet the difficulty is not insuperable. The first half of the clinical course in medicine, or some modification of this leading to a Master's degree, contains the elements of what is required. The problem is not to extend factual knowledge, but to increase receptivity to the significance of other categories of relevant experience.

I have chosen the situation in the biomedical field as an example of the problem because this is the field that I know best. I am satisfied, however, that a similar problem exists in many other provinces of natural knowledge and is equally in need of attention.

Up to this point we have been considering the qualitative aspects of the training of intending research workers. But there is also the quantitative.

Today, the needs of a country for research workers in all branches and at all levels of science are rising rapidly. If these are to be met, the means for training intending workers at the postgraduate level must increase correspondingly. At this level the requirements are essentially twofold: access to advanced facilities and intimate association with those who have established their position in research. During the last two decades there has been a substantial expansion in the provision for postgraduate work in university departments, and this has gone a considerable way to meet the need. Apart from the universities, however, there now exist many research institutes and units with excellent facilities and staffed by some of the most able and experienced research workers in the country. These are well suited to undertake training at the postgraduate level and many, by the orientation of their interests towards the fields of more specialized scientific knowledge, to extend the range of this. Clearly it is not to the advantage of any country that facilities of this quality and on this scale should not be fully used or that research workers who wish to teach should be unable to do so. Most postgraduates, however, are intent on obtaining a research doctorate, and as yet relaxation of the regulations governing the award of these by universities has not in general gone far enough for full advantage to be taken of the opportunity.[1]

VIII

The question that now remains for consideration is that of university policy in respect of research. Clearly, because of their unique responsibility for higher education, the universities must have a policy in respect of this. Can they, however, have in addition to this a policy for the development of research in its own right, or must research in universities be regarded as something incidental to their essential purpose of higher education? This is an important question both for the universities and for the society of which they are part. It deserves, therefore, our close attention.

Accepting that the primary purpose of the university is its unique responsibility for higher education, we may remind ourselves of the justification for associating research with this. Essentially this rests on the obligation of those who impart knowledge to verify that which they teach. This is a personal responsibility and as such constitutes the basis for the academic tradition of individual autonomy in intellectual matters. In any instance, therefore, the decision to initiate research and what research to do, within the ambit of the subject he professes, rests ultimately with the individual. This is the essential basis of all academic freedom and it is difficult to see how, consistently with

[1] The University of London awards two types of degree: the 'internal' and the 'external'. In respect of undergraduate degrees, the former are taken by students working within the University's own establishments; the latter by students working in appropriate places elsewhere. Postgraduate degrees, such as the 'External' PhD, have been open to any holder of a first degree of London University, like the senior doctorates of science and medicine, irrespective of where or with whom they work. Undoubtedly this traditional liberality of London University in regard to external degrees has been an important factor in equipping this country to face the scientific revolution. More recently in this country a body, the Council for National Academic Awards, has been created to help in this problem. It possesses no establishments of its own but awards degrees, including the PhD, to those working in suitable places other than universities.

this, a university could ever impose a policy for the development of particular research on the members of its staff.

It may be felt, however, that whilst this is quite true, a university could (like a central research organization) so select its staff that their individual interests would collectively allow a particular research policy to be realized. To do this, however, would entail the university in putting the interests of research before those of higher education. Thus, from the research point of view, it might be desirable to set up a hitherto unrepresented subject in a particular university, or substantially to expand a department that already existed; yet from the standpoint of educational policy, there may be no justification for this. Again, educational policy may require that a university so deploy its resources as to develop certain of its departments more than others and, as the volume of research in any particular subject generally reflects the number of academic staff concerned, thereby determine a balance of effort different from that which the interests of research would have required. It would seem inescapable, therefore, that an institution of higher education must always, in considering the deployment of its resources, put the requirements of education before those of research. In these circumstances it would be unrealistic to believe that a research policy in any significant sense could be implemented. Quite properly, the attitude of the university as a university must be that the promotion of research over and above that which follows from its educational policy is the responsibility of others.

But it is not possible to consider the university system and its relation to the over-all provision for scientific research entirely in terms of an individual university. In any scientifically developed country there are many universities and their collective significance for the development of its research also requires consideration.

By constitution all universities are separate autonomous bodies and each is free to further its own development to the best of its ability. In this country the existence of a University Grants Committee, charged with allocating the Government grant to the universities, has led to a measure of co-ordination arising in educational or administrative matters. But thanks to the Committee's insistence that allocations to individual universities shall be in the form of block grants, the essential principle of university autonomy has been preserved.

The University Grants Committee, like the individual university, has as its *raison d'être* the support and development of higher education. It is concerned with research only to the extent of taking into account, when making its block grants, the need for funds to be available to each university to support a basic level of research by its academic staff. Further than this the Committee could not go without involving itself in the potential conflict between educational and research requirements which, at the national level, would be many times more severe and productive of discord than it would be in the individual university. Today, research enjoys great prestige and its development to a high level in a university is a potent factor in recruitment and in the ability to attract financial support. Inevitably, therefore, in the field of research, individual universities are in competition. In these circumstances it would only be possible for the universities collectively to formulate a policy for research if all agreed to forego the possibilities of individual advantage now open to them. That the dead level of uniformity that would thereby result would be of advantage either to them or the country can hardly be maintained.

It seems, therefore, impossible on every count for the universities, either individually or collectively, to consider undertaking responsibilities for research policy without abandoning their essential nature and the principles that have brought them to their present high position in scientific development. And there is another and entirely different set of considerations that lead to the same conclusion.

Today, in any developed country, the demand for new scientific knowledge is insistent both in range and time. No country can now afford to depend entirely on the hope that the spontaneous curiosity of individuals will fortuitously fill its needs or that all the necessary co-operation between men in different disciplines will spring spontaneously into existence and be maintained under conditions of individual autonomy. Yet if today we were to require our universities to cover all our needs for research, in any province of knowledge, we could do so only by imposing on them such a measure of direction as would to all intents and purposes destroy their independence.[1] The existence of other instruments for scientific research with other responsibilities is, therefore, both a necessary complement to the functions of the university in research and a guarantee for their continued autonomy.

IX

We can now attempt to summarize the main points made in the preceding discussion in regard to the particular characteristics of the university considered as an instrument for the development of scientific research.

The whole function of a university and its particular value in respect of the contributions of its research to the over-all development of scientific knowledge derives from the principle of academic autonomy. By providing multiple centres of free enquiry, the universities constitute an indispensable national insurance against error, oversight and the persistence of wishful thinking. Because of the multiplicity of interests that are represented in each, they provide the kind of intellectually comprehensive environment that is conducive to the development of original work. Necessarily their prime responsibility for promoting higher education determines their policy in respect to the subjects they cover and the extent to which each is developed. Similarly, because of their association with higher education, the balance of research within them tends towards the unspecialized, and for such research the universities are favourably placed because of their relative insulation from other responsibilities. These are the great strengths of the universities in research. It is these, not deliberate

[1] 'In Great Britain where it is commonly supposed that the autonomy of the universities is perhaps most jealously guarded, scientists who call for larger government funds for science in the universities are prepared to employ arguments which, in effect, destroy the case for university autonomy.' J. Jewkes, D. Sawers and R. Stillerman: *The Sources of Invention*, p. 241. Macmillan & Co., London, 1958.

scientific policy, that in the conditions of the last century and a half brought the university to its present eminence. In the context of a national policy for the continued development of scientific knowledge, the universities are now, beyond question, a major and essential instrument. But they do not in themselves provide the whole context. There are other instruments, the functions of which are complementary, and of these the most important are the central organizations for scientific research that have come into being as a response to the growth of scientific knowledge and its increasing importance.

* 8 *

CENTRAL RESEARCH ORGANIZATIONS

I

TODAY, central organizations for the promotion of scientific research have developed in all modern countries. That this is in essence a natural response to a developing situation in scientific knowledge is suggested by the appearance, over a relatively short period of time, of such organizations in so many different places and by the similarities that these have shown in the stages of their evolution (Chapter 6). The needs that have led countries to take such action are not hard to seek. Scientific knowledge has become so important to society and is now drawing so substantially on public resources that some device for its informed support and for its integration into the structure of government had become a necessity. It is to the credit of this country to have pioneered in this development, and thereby to have set a pattern that in principle, if not always in form, has been widely followed.

The problem to be solved was not an easy one. On the one hand there is Government (or to be more precise, the different Departments of Government) with their need for information relevant to the questions of the moment. On the other are the scientific communities which, if they are to advance knowledge effectively, need to be substantially insulated from the changing distractions of practical affairs. On the one hand there is the administrator, or businessman, harassed by questions urgently demanding answers and seeing scientists engaged in work that seems to him essentially irrelevant. On the other is the scientific investigator, intent on wringing knowledge from the unknown, who sees in any attempt to persuade his interest an infringement of the conditions that are essential for his work as improper as it is misconceived. In principle, this latent conflict of purpose between the practical man and the expert is no new problem and the mutual suspicion to which it has given rise has been a fruitful source of failure in social organization over the centuries. It was, therefore, something of a landmark in the evolution of administrative thought to have produced a workable solution that reconciled these two potentially disruptive forces. This achievement will, therefore, repay some attention.

II

When, in the first and second decades in this century, the need arose to look at the provision for developing scientific knowledge in a national context, it was the singular good fortune of this country to have at the centre of its affairs three men who were exceptionally well qualified to appreciate the nature of the situation. These were the jurist and philosopher Haldane of Cloan, the ex-professor of anatomy and dean Christopher Addison, and that brilliant and unorthodox civil servant Robert Morant. Their solution to the problem was the independent Research Council standing half-way between Government and its Departments on the one hand and research workers and their institutions on the other and having executive authority to administer funds at its sole discretion subject to Parliamentary scrutiny of its transactions. As this device first reached its full development in the council concerned with biomedical research, the establishing of that organization and the significance of this can be taken as an example.

Although Government Departments had for some time made some provision for research on the medical matters that were of direct concern to them, more general provision for medical research arose only as a result of the National Health Insurance Act of 1911. In this it was agreed that for every person insured under the scheme the Treasury would make available, from public funds, an extra penny for medical research. In total this amounted to an annual sum of about £40,000. There being no Ministry of Health at that time, the Act was administered by Commissioners,

THE ORGANIZATION OF SCIENTIFIC DEVELOPMENT

and the organization to deal with research was made responsible to them. This consisted of a large representative advisory council (which fortunately proved to be a dead letter from the start), and a Medical Research Committee consisting of six scientific and three lay members with a whole-time medical executive secretary. In the interval between the passing of the Act and the setting up of the Committee there were two schools of thought as to the policy to be followed. One held that the Committee should concentrate on a few specific and pressing problems like tuberculosis; the other that it should adopt a wider and deeper policy of support for medical research as opportunity arose. Fortunately the latter view prevailed and its fruits were seen a year or two later when, within a short time of the outbreak of World War I, the Committee found itself deeply engaged in research problems for a diversity of previously unforeseen interests rather than being, as it might well have been, confined to problems of immediate concern to the strictly health field. Thus, in addition to the straight problems of illness in temperate climates, it was working on tropical medicine, the scientific basis of food rationing, wounds and wound infection, the physiological problems of flying and life in submarines, and with the Health of Munition Workers Committee on industrial poisoning in munition factories, accident proneness and sickness absenteeism. The significance of this catholic display of the relevance of medical research was not lost upon those who became concerned with future social planning.

It was at this stage (1918 to be exact) that the committee set up by Addison, with Haldane as chairman and Morant as a member, produced its classical report on the Machinery of Government.[1] In this they dealt thoroughly with the problem of scientific research and its relations to Government. They distinguished, in this context, two types of research. The first was 'Research Work supervised by Administrative Departments', by which they meant enquiries of the kind affecting the business of the Department that we should now perhaps call 'operational research'. This they

[1] *Report of the Machinery of Government Committee*, 1918, Chapter IV. Cd 9230. H.M.S.O., London.

saw clearly was the direct responsibility of the Department concerned. The second was 'Research Work for General Use', and they cited as examples of this the work of the Medical Research Committee and the more recently constituted Department of Scientific and Industrial Research (D.S.I.R.). This kind of research was specifically defined as that for the advancement of knowledge. Haldane then went on to consider the conditions required to make this possible. He points out that 'science ignores departmental as well as geographical areas'; that any organization concerned with general research must keep in touch with scientific workers of many fields and not only those where knowledge might be required for an immediate 'ad hoc' purpose; that enquiries started to answer a short-term practical question must often be carried through into the long-term and not relinquished, as an Administrative Department must needs do, if they go deeper than the immediate purpose; and that a general research organization must not be weighted with continuing responsibility for implementing any recommendations that arose out of its research findings. On the basis of these considerations the Haldane Committee then laid down their two famous principles. First, that research in the sense of enquiry into naturally occurring phenomena should be set up, and established, independently of the administrative departments concerned with its findings. Second, that operational research in the sense of work affecting the performance of a particular administrative Department (and, we might add, industry) should be the ordinary practice and responsibility of such Departments. It thus came about that the scientific research councils were set up under the Privy Council[1] and not under Departmental Ministers.

The test of this doctrine followed swiftly. In the next year, Addison was commissioned to pilot a Bill through Parliament for the purpose of setting up a Ministry of Health. In agreement with the recommendations of the Haldane report[2] it was proposed

---

[1] The Minister responsible for the Privy Council would, in other countries, be regarded essentially as a 'Minister without portfolio'.
[2] *Report of the Machinery of Government Committee*, 1918, Chapter IX, para. 15.

that the old Medical Research Committee should not be taken over as the research department of the new Ministry but be set up independently of it. In the debate on the Second Reading, Addison came under strong pressure to withdraw this proposal and to put the Research Committee under the direct control of the Minister of Health. In reply he put forward a memorandum that is second in fame only to the Haldane report.[1] After marshalling a series of cogent objective reasons why his proposal should prevail, he came to the nub of the matter which, as Minister of Health designate, he was able to put forward with particular effect. Now that scientific knowledge has become of such, and on occasions, literally vital, importance to a society, this point involves a principle of such wide-ranging importance that it is worth giving his argument in his own words, although he was speaking only in the context of health:

'A progressive Ministry of Health must necessarily become committed from time to time to particular systems of health administration. . . . One does not wish to attach too much importance to the possibility that a particular Minister may hold strong personal views on particular questions of medical science or its application in practice; but even apart from special difficulties of this kind, which cannot be left out of account, a keen and energetic Minister will quite properly do his best to maintain the administrative policy which he finds existing in his Department, or imposes on his Department during his term of office. He would, therefore, be constantly tempted to endeavour in various ways to secure that the conclusions reached by organized research under any scientific body, such as the Medical Research Committee, which was substantially under his control, should not suggest that his administrative policy might require alteration. . . . It is essential that such a situation should not be allowed to arise, for it is the first object of scientific research of all kinds to make new discoveries, and these discoveries are bound to correct the con-

[1] *Memorandum on the Ministry of Health Bill, 1919, as to the Work of the Medical Research Committee.* 1919, Cmd 69. H.M.S.O., London. The whole of this classic memorandum is worth perusal. The quotation is from paragraphs 10 and 11.

clusions based upon the knowledge that was previously available and, therefore, in the long run to make it right to alter administrative policy.... This can only be secured by making the connexion between the administrative Departments concerned, for example, with medicine and public health, and the research bodies whose work touches on the same subjects, as elastic as possible, and by refraining from putting scientific bodies in any way under the direct control of Ministers responsible for the administration of health matters.'

Despite the unconventional nature of his proposal, Addison prevailed and in so doing secured for this country two essential requirements for the new age into which it was entering. For Parliament and public alike he secured a continuing source of expert opinion which, because of its independence of commitment to administrative policies and sectional interests, was demonstrably impartial. For research workers he provided a type of organization which, being scientific, received their professional confidence and thereby enabled the creative ability of the country to be mobilized.

It is salutary to consider what would have happened had Addison failed. Parliament and country would thereby have been deprived of access to a permanent source of informed opinion on matters of acute political concern, such as the hazards of 'fall-out' from atomic explosions, the safety of new vaccines, the rationing of food in war-time, the risks of economically important habits like cigarette smoking and the dangers of particular industrial operations, which by its independence of the interests involved could be generally accepted as disinterested and unselected. There would have been no bridge between the academic and research communities on the one side and the world of practical affairs on the other, to the detriment not only of the growth of the desired co-operation, but also of the development of knowledge in those fields in which their respective experience was mutually complementary. But Addison did not fail, and his vindication came nearly half a century later in the conclusions of the 'Committee of Enquiry into the Organization of Civil Science':

'The Haldane Report established the principle that the control of research should be separated from the executive function of Government. The Research Council system, as we know it today, is based on this principle which, in our view, has contributed significantly to the Council's ability to promote research and development while simultaneously guaranteeing the independence of the scientific judgments involved. We endorse this concept that public funds provided for the support of scientific civil research should be administered by autonomous Research Councils; and we consider that such Councils should continue to be responsible for much of the scientific research sponsored by Government.'[1]

This, then, the concept of an autonomous independent Research Council standing between Government and research, was the supplementary device that evolved in the second decade of this century in response to the needs of the new situation created by the rapidly developing importance of scientific knowledge. It is evident that this concept has proved applicable to the biomedical province of knowledge. The question is, however, whether it is similarly applicable to other provinces.

III

The continued success of any human organization depends upon its satisfying both of two basic requirements. The first is that it shall be in conformity with the deep-rooted sentiments of the men who have to make it work and thus productive of the requisite loyalty and morale. In relation to the device of a research council this was secured by making the new bodies autonomous and executive and by putting them under the professional control of scientists themselves. The second is that it shall be equally in conformity with the natural realities with which it has to deal.[2] It is in this connexion that our ideas on the structure of scientific knowledge assume such importance, for according to the image we have of this so shall we devise our

[1] *Committee of Enquiry into the Organization of Civil Science*, 1963, para. 61. Cmnd 2171. H.M.S.O., London.
[2] James Bryce: *The Holy Roman Empire*, Chapter XXIV, p. 488. Macmillan, London (Papermac 193).

organization for its future development. Thus, for example, if we are imbued with the concept of a growing tree of knowledge, it might appear eminently reasonable to us to construct one central organization to cover all scientific research, with subordinate branches corresponding to different activities. If on the other hand we believe that scientific knowledge has in fact developed from many different points and that the situation is more accurately represented as a series of adjoining provinces of knowledge arranged round the surface of a vast globe of ignorance then we shall clearly incline towards a confederation of organizations each of which corresponds to a province of natural knowledge. I have given my reasons for believing that this latter concept is the more consistent with the reality. If this be accepted, however, it has an important implication. It is that, at the national level, the unit of organization is one that corresponds not to any subject, however large, nor to any quality of research (such as 'pure', or 'applied'), but to one of the provinces of natural knowledge which ranges from specialized knowledge at the periphery, where it abuts on advanced practice, to unspecialized nearer the centre, where adjoining provinces are tending increasingly to merge. We might, therefore, look at the system of research councils from this point of view.

As a result of the reforms instituted in the second decade of this century, three research councils came into being in this country. These were the Department of Scientific and Industrial Research, the Medical Research Council and later the Agricultural Research Council. That these not only survived but developed progressively over the next forty years is a substantial indication of the soundness of the concepts upon which they had been based. At the end of that time, however, it was felt that the whole situation should be reviewed. Accordingly a governmental committee, under the chairmanship of Sir Burke Trend, was appointed early in 1962 to consider the whole question of the organization of civil science supported from public funds. This reported at the end of 1963 and their findings are of great significance not only in this country but also in others that might be influenced by our example.

THE ORGANIZATION OF SCIENTIFIC DEVELOPMENT

The Trend Committee endorsed, without reservation, the Haldane principle that the control of scientific research should be separated from the executive functions of Government and that support for scientific research from public funds should be administered by autonomous research councils. Despite this endorsement, however, the Committee felt bound to suggest alterations in existing arrangements; and for our immediate purpose, one of these was of great significance. Whilst recommending no essential change in regard to the scope and activities of the Medical and Agricultural Research Councils, the Committee advised that the body responsible for research in relation to industry, the Department of Scientific and Industrial Research (D.S.I.R.), should be dissolved and its functions distributed between new organizations. Clearly such a drastic recommendation needed full justification, and that advanced by the Committee was that the responsibilities which, under existing arrangements, the D.S.I.R. was required to meet, would become too heavy a charge for one organization.[1] But was the difference between the responsibilities of the councils concerned with medicine and agriculture on the one hand and that related to industry on the other entirely quantitative, or was it more fundamental? Putting it another way, was there a defect in the organizational concept of a scientific and industrial research council which the other two research councils had escaped? This question is well worth considering, for it can hardly fail to reveal lessons for the future.

The Department of Scientific and Industrial Research (D.S.I.R.) was established in 1916, two years before the Haldane Committee reported and before the concept of a research council had been clearly formulated. Its recognized purpose was 'to promote and organize scientific research with a view to its application to trade and industry'.[2] From the outset, however, it laboured under a particular handicap. In both the medical and

[1] *Committee of Enquiry into the Organization of Civil Science*, 1963, para. 65. Cmnd 2171. H.M.S.O., London.
[2] *Report of the Machinery of Government Committee*, 1918. Chapter IV, para. 45.

agricultural fields, the division of responsibility for operational research and for research for general purposes, recommended by Haldane, had been put into effect, the appropriate Ministry being responsible for the former, the appropriate research council for the latter. There was thus a sufficient division of responsibility for each to be able to frame its policy consistently, and acting together, form an effective partnership. In the case of the D.S.I.R. however, no corresponding Ministry existed. Further, such operational research relating to industry as existed was scattered throughout industry itself, under the control of particular industries or even individual interests within these. The effective partner, responsible for operational research, was thus lacking and, in default of this, D.S.I.R. was impelled to meet the need itself. In an attempt to remedy the situation, individual industries were encouraged to set up Research Associations towards which the D.S.I.R. gave substantial financial assistance. In addition, the D.S.I.R. supported research stations of its own concerned with specific needs, or, as in the case of the National Physical Laboratory, to undertake work of wider relevance. Thus, from the outset, D.S.I.R. was driven increasingly to contravene one of the principles that the Haldane Committee had considered essential, namely that a central research organization should not be required to assume continuing responsibility for ensuring the application to practice of the recommendations that arose out of its relevant research findings.[1]

The second difficulty arose from the very width and diversity of industry itself. Some indication of what this meant for D.S.I.R. is given by the fact that, at the time of its dissolution, it was concerned with fifty-three different Research Associations each related to a different industry and together representing practically the whole variety of industrial interests. Clearly even to keep in being such a width of responsibility was to impose a heavy preoccupation on any organization. To expect intense attention to be given to all particular major problems as they arose, was to ask too much. In these circumstances, it came about that certain major new developments, and the more specialized

[1] *Ibid.* Chapter IV, para. 63.

research related to these, such as that in the field of atomic power for civil purposes, were not included in D.S.I.R. but made the responsibility of separate organizations.

But this was not all. In the course of its evolution D.S.I.R. had become, in addition, the major external source for the support both of unspecialized research and postgraduate training programmes in the universities, not only in those scientific subjects related to industry, but also in regard to unspecialized research in general.

On any count, this aggregation of responsibilities was an impossibly wide and indeterminate load to place on any organization and it reflects the highest credit on the men involved that they coped to the great extent that they did. It is no wonder, however, that the Trend Committee felt that the situation that had evolved called for attention. But what exactly were the factors in the overload?

The particular overloading which the Trend Committee felt justified their recommendation to dissolve the D.S.I.R. was that which, in their opinion, resulted from the attempt to combine responsibility for promoting research in universities with that for promoting industrial research and development.[1] If this diagnosis were correct, then it clearly pointed to the remedy. The two responsibilities should be separated and an organization designed for each. This was in fact the recommendation that was made. An authority was to be created to assume responsibility for the research arising out of the specialized experiences of industry; a Science Research Council was to be formed to promote unspecialized research in the universities. Although somewhat modified in the course of implementation, this was substantially what came about. But there was an alternative diagnosis: that the overloading arose not from the attempt to combine responsibility for promoting the whole range of relevant knowledge from the specialized to the unspecialized (for which purpose the D.S.I.R., like other research councils, had been originally devised), but from the inordinate width of the fields that the D.S.I.R. was being expected to cover; that in effect, the situation was that a single

[1] *Committee of Enquiry into the Organization of Civil Science*, 1963, para. 65.

research council was being asked to cover, not a single province of knowledge, but a whole continent of different provinces. In retrospect this, in my opinion, would have been the correct diagnosis. Had it been made, the remedy proposed would almost certainly have been different. It would have been along the lines of dividing up the industrially orientated field on the basis of natural provinces of scientific knowledge (just as biomedicine and bioagriculture had been distinguished in the biological field), and constructing a separate research council for each.[1]

The question is, was this possible? Clearly, if there were no such provinces in the industrially orientated field, it would have been merely an expedient to fabricate them and no benefit could have been expected to come from legislating on these lines. But is this, in fact, the case? The history of the development of scientific knowledge and the dynamic interplay between specialized and unspecialized research that has been a feature of scientific progress in all its stages, suggests that this might not be so. It is, therefore, worth looking at this particular field to see if there are any indications of natural divisions within it.

I think that, in approaching this intransigent problem, we should keep firmly in mind that the primary purpose of scientific research is to increase our understanding of the natural world and that it has its origins in our endeavours to extend our ability to satisfy certain perennial human needs. In respect of those penetrations of enquiry that have ultimately given us both modern industry and the physical sciences, it is possible to identify five such needs. These are the needs for materials, energy, transportation, communication of information, and structures. May we now glance at each of these?

We have already considered the sequences of necessary knowledges that led down from colours, metals, fabrics, drugs, natural

[1] At the time that I, personally, gave evidence to the Trend Committee, I was still largely groping towards the concept of provinces of knowledge. Like the Committee I felt that the remit of the D.S.I.R. was intolerably wide, but my suggestion was then to amputate the Research Associations, hand those to an appropriate Government Department, and otherwise leave D.S.I.R. as it was. The revised diagnosis that I now put forward must, therefore, be regarded largely as a subsequent development of my own thinking.

products and so on to the common stratum of unspecialized knowledge and theory that we call chemistry. Together these form a province of natural knowledge in respect of materials. We have similarly considered the sequences that led down from our endeavours to harness naturally occurring forces, molecular and atomic energy, into a common stratum of physical knowledge and theory (including nuclear physics) and which, by their dependence on the invention of the appropriate machines, have brought into existence a variety of engineering disciplines. These in their turn together form a natural province of knowledge of energy. Transportation, with its origins in the problems of moving men and materials on land, on sea, in the air and perhaps in space, we have touched upon. These sequences lead down through specialized aspects of structural theory and dynamics to a particular emphasis in general physical knowledge—and like the previous province, this also has because of its dependence on machines given rise to special disciplines of engineering. This province is further closely contiguous with the previous one concerned with energy (rather as bioagriculture and veterinary medicine are contiguous with biomedicine) and also with communications; and, like these, it touches on psychological considerations of the human factor. The nucleus of the suggested province of communications knowledge is almost in existence already in the span of research in the Post Office, telecommunications and the contacts that these have established with relevant knowledge in subjects as far apart as engineering and psychology. This might also be given the main responsibility for computer science. The question might well be asked at this stage, where do you suggest that subjects like electronics and space research find their place? Just as the stratum of biochemistry traverses the provinces in the biological field and penetrates into the province of materials, so electronics and solid state physics spread across several provinces related to physics. Space research is more difficult, but at present it would seem to belong most naturally in the province of communication of knowledge. That brings us to the fifth of the suggested provinces of knowledge, that of structures in the sense of buildings. This derives from architecture in the broad sense of the term. In this sense it covers not only the theory of structures, materials and design, but also the arrangement of buildings in town planning and the physiological and psychological requirements for their effective utilization.[1]

These then are the kinds of suggestion that I would put forward as a basis for organizing scientific knowledge that finds its expression, and draws more inspiration than is often appreciated, from the problems of industry. I advance them with considerable diffidence, not because I doubt that intellectual provinces exist in this part of scientific knowledge, but because I am very conscious of my relative lack of qualification to identify them. In suggesting them, however, I am fortified by one consideration. The groupings that I have suggested are to a considerable extent already indicated in the structure we give to our educational curricula.

It is a sound precept in planning a course of education to proceed from the general to the particular. Take medical education, for example. The student normally enters on this at the unspecialized levels of chemistry, physics and biology. Thereafter he proceeds to the study of the normal: human anatomy, physiology and biochemistry. He then moves towards the abnormal through pharmacology, microbiology and pathology until he reaches the clinical fields of medicine, surgery, obstetrics and so forth. His course has been so constructed that he passes from the unspecialized to the specialized, and in passing through it he has in fact retraced in the opposite direction the path that knowledge took in the process of its development. At each step in the course he is using previous knowledge that originally was sought to elucidate problems at the level with which he is now engaged. His teachers, and their subjects, are arranged in the same order, and at any stage in the course he, or they, may decide to stop and not continue to the specialized periphery. Naturally, if it has been the intention from the outset to proceed only a certain distance from the unspecialized beginning, say to biochemistry, and courses have been designed specifically for people with this intention, it will not be so evident that the particular knowledge is

[1] Llewelyn-Davies and P. Cowan: 'The Future of Research,' *Journal of the Royal Institute of British Architects*, April 1964.

in fact the unspecialized part of a lengthier intellectual sequence. If, therefore, attention is confined to curricula designed to produce pure chemists or pure physicists, one might not easily appreciate that these were, in essence, the unspecialized part of an intellectual sequence. If however one looks at curricula designed to train people for the specialized periphery of scientific knowledge, as in medicine, agriculture and the various branches of engineering, geology, architecture and so on, there is an evident indication that, consciously or not, the basis upon which these have been constructed is that of a province of natural knowledge.

IV

To complete the sketch of the present situation, two further provinces require consideration. The first concerns natural resources and the second the problems of men living in organized societies.

Of necessity, men are dependent for the raw materials they use on their knowledge of the natural environment. As we have seen, their endeavours in this respect initiated the development of the 'earth sciences' and together these appear to form a coherent province of natural knowledge. Now that the demand for materials is rising, when the more accessible sources are being rapidly exhausted and when we are depleting our planet to the extent that we are, the need for a scientific organization that can assess objectively and independently the consequences of our actions and indicate the measures required to meet our needs has become indisputable. But natural resources are all part of our environment and it is only in the context of the totality of considerations that bear on this, that they can be understood. It was, therefore, a wise provision of the Trend Committee in this country to assemble the scattered studies in this province and bring them together into one central research organization. These together with meteorology now form a natural environmental research council. Whether astronomy falls naturally within the same intellectual remit as the earth sciences, or whether it is more appropriately related to the province of communications knowledge, together with space research, is a matter for expert consideration.

There remains the problem of sociological studies. The urgent need for scientific knowledge about human societies makes, even in the present rudimentary state of development of many subjects in the field, organization for the promotion of knowledge in the province of sociology a pressing necessity. This is fraught with particular difficulty for in it we are concerned not only with natural phenomena but also with epiphenomena such as legal and economic systems that men have devised empirically to enable them to cope with the problems of social living. Nevertheless, we cannot doubt that the natural phenomena themselves are susceptible to study scientifically, nor that provision should be made for the integrated development of this province, or perhaps continent, of knowledge. Perhaps, at this present stage, cultivation of those sequences that adjoin the biological provinces may provide the surest line of advance. In this country, at least the beginning has been made.

V

This then is the position as I see it. Administration expresses itself through the organizations that it devises. At the beginning of this section it was laid down as axiomatic that the success or failure of any human organization depended on the extent to which it satisfied both of two requirements. The first was that it should be in conformity with the deeply held sentiments of those who have to operate it, for only thus could it engender morale. The second was that if it were not to be merely a temporary and unsatisfactory expedient, it must conform to the natural realities underlying the problem with which it sought to deal. The thesis here put forward is that the natural reality underlying the concept of a central research organization is that of a province of natural knowledge extending from the specialized periphery that abuts on advanced practice to the unspecialized regions towards the centre where all the provinces tend together. I am satisfied that such natural provinces exist and that it is on the basis of these that scientific knowledge has developed. Until the growth of knowledge forced specialization and its consequent restriction of interest upon us, it was in contexts of such breadth that individual scientists operated. Now this is becoming increasingly more difficult. Yet now, as always, progress depends on maintaining the intellectual unity that underlies the diversity of knowledge. That is the essential function of a central research organization, and from this all its other functions flow. The reality to which the different central research organizations are required to conform is, therefore, that of a natural province of scientific knowledge. In default of that, they have no *raison d'être*, and only in so far as such exists and they conform to it can they be effective.

We have considered the relationship of research in universities to the development of knowledge in these different provinces of science. Before considering central research organization further, it is necessary to consider the contribution of research arising at the level of practice and the problem of policy for the development of scientific knowledge.

9

RESEARCH IN THE PROFESSIONS AND INDUSTRY

I

IN A SENSE, research in relation to practice, in that it has as a major aim the translation of scientific knowledge into practical achievement, provides the means through which scientific research achieves its ultimate justification in the eyes of society. Without this, the importance of scientific knowledge would fall to be judged purely in terms of its cultural interest, and, great as this is, it is unlikely that, by itself, this would suffice to obtain for it resources on the scale now required for its development. But leaving such stark considerations on one side, and leaving aside also any question of the moral obligations that scientists incur because of the special knowledge that they have been enabled to acquire, there is another and purely intellectual reason why the proper support of research at the level of practice is an essential requirement in any over-all provision for the development of scientific knowledge.

Research arising from practical considerations is necessarily primarily concerned with problems at the interface between advanced practice and the external frontier of the province of knowledge to which it is related. It is thus on the one hand continuously exposed to the salutary lessons of unpremeditated experience and on the other to relevant developments in scientific knowledge. In consequence it is a fertile source of unexpected data which, more frequently than is often appreciated, stimulate trains of thought at other and more theoretical levels. Further, because of its access to the wide diversity of situations in practice, and its understanding of these, research at this level provides the means by which ideas evolved in more abstract studies can be subjected to extended and exacting verification. These are indispensable contributions. Without them, thought at the more unspecialized levels of research would be in danger of degenerating into a form of scholasticism and knowledge over the province as a whole to lose perspective. To regard such research as concerned only with the application of scientific knowledge to practical affairs is, therefore, to overlook a large part of its significance. To provide for its proper support and the integration of its contributions with other knowledge must, therefore, be an essential task of any national policy for the development of scientific knowledge.

II

The problem of promoting research in association with practice necessarily varies with the practice in question. If one is concerned, as one is in medicine, with a profession in which there is an established tradition of co-operation, there is in principle little difficulty. If, on the other hand, one is dealing with industry, where the tradition is competition, the problem is essentially different. In both cases, however, the situation is one in which the potential sources of initiative are multiple, each of which is entirely free to decide whether or not to undertake research and the extent to which it will co-operate with others should they decide to do so. In essence, therefore, any policy for the promotion of research at the level of practice must rely largely on the agency of local initiative and yet, at the same time, ensure that the dispersion of effort this entails is effectively related to the body of scientific knowledge on which it draws and to which it contributes. To meet the first requirement, responsibility for the research will need to be largely decentralized; to meet the second, a strong central research organization is required. As an illustration of a form of arrangement that has proved suitable for research in relation to one kind of practice, that for clinical research[1] in this country may be cited.

[1] By 'clinical research' is meant research that is directly concerned with the problems of illness in human beings.

RESEARCH IN THE PROFESSIONS AND INDUSTRY

Clinical research is supported through two separate channels: at the decentralized level by hospital authorities, centrally by the Medical Research Council on the advice of its Clinical Research Board. The kind of research supported by the former agencies under this particular scheme is described as that which arises out of and is inseparable from good practice. Essentially support at this level provides grants for assistance of limited duration and is excluded from supporting major long-term research programmes involving the services of men who have elected to make their careers in whole-time investigative work. Decentralized clinical research of this kind is financed by direct block grants from the Health Department of Government and augmented by local private funds. These resources are entirely at the disposal of the local agencies and, being disbursed on the advice of local research committees, they have proved an effective stimulus for local initiative. At the central level the concern is with major projects requiring long-term support, the provision of whole-time careers and the integration of specialized clinical research with research at less specialized levels. Nevertheless it is open to any medical man to apply to the central agency for support for projects of the kind normally supported under the decentralized scheme if he is not satisfied with the local assessment.

These arrangements have worked well. But medicine is in a peculiarly favourable position. It is staffed by a body of men brought up in the tradition that the advancement of knowledge is a professional duty. It is liberally supplied with professional journals which are widely read as a matter of course. Knowledge is freely communicated and its concealment unthinkable.

This, however, is only part of the picture. Although practice operates at the level of individuals, it takes place in communities and it is often only at the community level that its achievements can be seen in perspective or certain kinds of information, essential for its purpose, can be obtained. Thus, both for the purposes of health administration and for those of medical research, it is necessary that reliable information should be steadily forthcoming on the death rates and illness rates over the community as a whole and in the different sexes or age groups, the incidence of different diseases and so on. Epidemics, due to infection or otherwise, are a major concern. Only at the community level can they be detected and the associated conditions that provide clues to their causation identified. Powerful drugs are increasingly coming into use, and although these are subjected to careful safety tests before being generally released, those with infrequent toxic effects are only likely to be detected by assembling information over the country as a whole. Again, at community levels below the national, at the level of particular occupational groups and skills, a comparable situation exists. The problems of health in different industries, those of the military services, those consequent upon the special habits of modern life like rapid transport or the processing of food, and many of those related to social insurance schemes, must largely be approached in the mass and with the co-operation of men who understand both the medical and related aspects. These are the territories of experience the needs of which Haldane, with his customary prescience, foresaw when he laid down his second principle that the responsibility for obtaining 'intelligence and information' should rest firmly with the executive Department of Government concerned. Without such, the community could not function. And equally without the background of information and experience thereby gained, the development of scientific knowledge in the relevant fields would be substantially handicapped.

In the relatively simple situation at the level of practice in the biomedical field, we can distinguish, therefore, three components in the over-all provision requiring to be made for translating scientific knowledge into practice and for the feedback of ideas and information from practice into the body of scientific knowledge. These are decentralized support of research at the level of practice itself, an appropriate central research organization covering the biomedical province of knowledge from its specialized to its unspecialized regions, and a series of governmental departments responsible for administrative action involving medical considerations in their different fields. Taking this as a prototype, we can now look at the much more difficult situation in the industrial field.

RESEARCH IN THE PROFESSIONS AND INDUSTRY

III

On analysis, the difficulty of providing for research in industry seems to be in part inherent and in part due to shortcomings in the administrative measures that have been devised to support it. The former derives from the competitive nature of industrial activities imposing a limitation on the degree of co-operation that can be counted upon; the latter from a failure as yet to devise an appropriate system of central research organizations with which industry can form links.

The competitive element in the situation must, necessarily, be accepted as in the nature of the case. In consequence it must also be accepted that the decision as to whether or not research is to be undertaken in any particular case will rest with the individual industrial firm or corporation. Essentially, the grounds for this decision will be economic. Only if the economic advantages promise to outweigh the cost of the research will this be done. If, therefore, the intention is to stimulate more research at the level of industrial practice, the first requisite is to ensure that the basic level of the financial encouragement to exercise initiative is readily available throughout the vast number of individual units of which industry is made up. As in the case of medical research at this level support must be decentralized. The most feasible way of doing this would seem to be through generous financial concessions in respect of the cost incurred in undertaking research.

To generalize about the units of industrial organization would be misleading. Today, these range in size from mammoth corporations to small businesses. The former are quite able to develop substantial research for their own purposes. The latter, however, are too small to do this effectively. It was to meet this need that the former Department of Scientific and Industrial Research sponsored the scheme of Research Associations whereby firms in a particular trade could join together to promote research of common interest or from which individual firms could commission work of particular and confidential interest to themselves. The number of Research Associations that were created is some indication of the size of this need.

Given a basic level of financial incentive for research throughout industry, and a means by which the smaller firms in the trades that are organized on this basis can achieve a basic level of co-operation, the groundwork would have been laid. On the basis of this, selective grants or contracts from the centre could be deployed with some assurance of effect. In essence, in this country, the groundwork is there. It only needs developing. But it is evident that by itself this is not enough. Economic considerations demand nowadays that the Government should have a close concern with industrial policy. Its selective intervention on a massive scale for its further development has become a necessity. Further, if scientific knowledge as such is eventually to find its expression in practical achievement, a means to relate this to practice is required. To date, however, these requirements seem largely to have defeated the attempts of any country at rationalization.

In the previous chapter reference was made to the report of the Committee on the Organization of Civil Science in this country and the considerations that led up to their recommendation that the Department of Scientific and Industrial Research should be dissolved. It is of interest now to look at their suggestions regarding the type of central organization that should take over the responsibilities for research in relation to industry that, in consequence of their recommendation, had now to be catered for.

After careful consideration the Committee decided definitely against putting research of this kind within a Government Department. Instead they recommended an independent organization (the Industrial Research and Development Authority) more akin to a research council but, in view of its heavy responsibilities for development work in partnership with industry, with larger freedom of action in executive matters. The new authority was to stand in the same relation to Government as the existing autonomous research councils. Its basic functions were to promote research and development within industry and to ensure that the universities and colleges of advanced technology were closely associated with its work. To this end it was to take over the research stations of the D.S.I.R., become responsible for the Research Associations, and provide grants to the universities, colleges of technology and similar institutions in support of research relevant to industry. Further, it would also absorb the National Research and Development Corporation that had been established previously to develop inventions and discoveries that had reached the patentable stage.

But would this recommendation, if accepted, have met the essential requirements? By the relative divorce of the new authority from concern with the more unspecialized research in the universities, it weakened the degree of integration in these fields of scientific knowledge. By making the new authority responsible for fostering research relevant to practically the whole range of industry, it perpetuated the situation that had, in my view, been primarily responsible for overstraining the Department that it was to replace. Had the Committee felt able to divide the field of responsibility, as has been suggested, in accordance with natural provinces of scientific knowledge included in this range, it is possible that they would have come to the conclusion that, in principle, the concept of a remit ranging from the specialized to the unspecialized, upon which the D.S.I.R. had originally been constructed, was a sound one and that the solution was not to discard this but to construct five or so miniature D.S.I.R.s on this basis.

The lines of organization that I see as appropriate for the promotion of research in relation to industrial practice are, therefore, similar in principle to those that exist in the biomedical field. Decentralization of support to the level of local initiative by means of financial concessions in recognition of the undertaking of research. The device of Research Associations for the trades that operate on the basis of small firms. A series of central research organizations, each corresponding to a province of natural knowledge and each relating both to the research of industrial practice and that of the universities and also to a series of Departments of Government with their separate administrative interests in the industrial field.

Given these components, a comprehensive scheme can be created both for the promotion and development of research in

THE ORGANIZATION OF SCIENTIFIC DEVELOPMENT

relation to industry and also for integrating the contributions of such research into the general body of scientific knowledge.

There remains, however, one question that must be raised in this connexion. That concerns the problem of industrial developments that can only be carried out on a massive scale. Can such projects ever profitably be related to a central research organization system, or must one always set up an entirely new organization that shall subsume in itself all necessary research, development and practice? To take a specific example. We have seen that when, in the aftermath of war, provision had to be made for the future development of atomic energy in this country, it was felt that the civil research aspects of this problem could not be assigned to D.S.I.R. but that an entirely new organization, the Atomic Energy Authority, should be created and include this purpose. In the circumstances, it is difficult to see what else could have been done. D.S.I.R., with its relation to practically the whole range and diversity of industry and its heavy responsibilities for promoting research in universities, was stretched to capacity. Further, there was no Government Department nor public or private corporation to which D.S.I.R., if it had undertaken to promote the necessary research, could relate in respect of the more operational research and development. If, however, the situation had been different, decisions might have been otherwise. If there had been a central research organization concerned with energy as such (and so able to devote the necessary intensity of attention and resources to the problem), and if there had been also a Government Department or public industry concerned with the practical development of atomic energy as a source of power, then it might have been possible to contemplate another arrangement. The less specialized research, upon which future development would depend, could then have been promoted independently of the user and in the context of research respecting other sources of energy, whilst the relevant Department or Industry would have developed the more specialized and operational research in relation to practice. Circumstances could well arise in which a country might benefit greatly from having available such an independent focus of knowledge which, by its close relation to the frontiers of research on all sources of energy, might be in a favourable position to give a comprehensive assessment of the scientific possibilities of one relative to another. Further, by linking research specifically orientated to the problems of the industry in question with unspecialized research in universities and research institutions, such a central research organization could be a potent influence in preventing the separation of one from the other.

In my opinion, therefore, size alone does not invalidate the principles of organization that are applicable to less enormous problems. Separation of the two extremes of a sequence or province of scientific knowledge is always to be deprecated. In industry no less than in other fields, long-term progress depends on the preservation of the continuum. It is the particular function of a central research organization to do this and, to this end, to mobilize appropriately all relevant talent irrespective of where it is situated.

* 10 *

RESEARCH POLICY AND PROVINCES OF KNOWLEDGE

I

IN GENERAL, it is a principle of organization for any purpose that only in so far as a field of experience is in itself coherent is it possible to formulate effective policy for its development. In respect of the development of scientific knowledge the first requisite is, therefore, to identify the fields of natural experience with which we have to deal and then to devise our organization to correspond with these. The central thesis of this essay is that our understanding of natural phenomena is built up on the basis of provinces of natural knowledge, each of which is specialized and mission-orientated at its periphery and all of which tend progressively to coalesce as they develop towards the centre where knowledge is more unspecialized and increasingly self-orientated. Accordingly it is proposed to approach the problem of providing for scientific development from the point of view that such intellectual provinces reflect the primary groupings of natural experience to which our organization, in institutional terms, must correspond.

II

Among scientists opinions vary widely as to the extent, if any, to which it is possible to formulate a positive policy for the development of scientific knowledge. In the past, it is said, we have managed very well without this. Research by its persistent probing has steadily found its way through an increasing number of problems, and, by the progressive aggregation of such tactical advances, the true direction of development reveals itself. Further, the history of the sciences is littered with examples of how progress has been misdirected by wishful thinking. Are we not, it is asked, in seeking after positive policy, following a mere will-o'-the-wisp? In opposition to this it is contended that, although the policy of *laissez-faire* may have been all that was possible in the previous state of our ignorance, that is not the position today. In recent years we have seen an increasing number of instances in which, by conscious thought, intended objectives have been achieved with a speed and certainty that establishes beyond doubt that planning is now feasible. Further, it is pointed out that the present demands of research on the national economies are such that choice is, in any case, inescapable.

Clearly unless some general agreement can be reached as to whether or not policy for scientific development is possible, advice from the scientific community will be conflicting, and the danger will be that, under the pressure of events, deleterious decisions will be taken. Leaving aside for the moment the question whether policy is feasible in respect of scientific knowledge as a whole, I propose first, therefore, to look at this problem from the standpoint of particular provinces of knowledge. Before doing this, however, it will be as well to deal with a prevalent misconception that often compromises discussion.

Among scientists there is a widespread fear that policy inevitably implies that they will be required to submit to direction in regard to the interests they shall pursue. Such fears are misplaced. Nothing but basic ignorance could excuse a belief that it is possible to require a creative worker to change his interest to order, and no competent research organization would ever entertain the idea. In respect of research, policy must express itself 'not through prescription to individuals but through the informed selection of projects to be supported'.[1] Faced with a need in research, the task is to find a man, or men, whose interests match this. Given such, policy realizes itself. In regard to the implementation of research policy, therefore, this is the governing consideration, and it is only on this basis that the problem can be approached.

[1] Medical Research Council, Annual Report, 1960–1, p. 8.

The objectives of research policy can be considered at several levels. At the most general, statements about these are almost of the nature of platitudes. None, for example, is likely to disagree about the desirability of our learning how to prevent or cure cancer, or to produce sufficient food in the world to allay hunger, or to have at our disposal illimitable supplies of cheap controllable energy. But, if such general statements are to graduate from the status of a wish to that of a purpose, it must first be shown that there are reasonable grounds for regarding their attainment as feasible. It is here, in relation to feasibility, that research policy begins. Given, therefore, broad agreement that certain objectives are desirable, it is a purely scientific matter to decide the extent to which they are feasible and the policy to be pursued to attain the end in question.

In relation to feasibility studies, several considerations come into account. First and of overriding importance is an assessment of the existing state of knowledge and the identification and close analysis of the inadequacies in this that have prevented its further progress. For example, it is evident in respect of malignant disease, as contrasted with microbial infections, that we are concerned not with the continued attack on the normal body by an outside agent, but with a self-sustaining abnormality of the cells of the body itself. Until, therefore, we know considerably more than we do at present about the controlling mechanisms that keep cells normal, we cannot hope to understand the situation. The progress of research in subjects like molecular biology is therefore of great moment. The second consideration is that of relative importance. Analysis may have indicated that progress in a particular province of knowledge is held up by intellectual inadequacies at a particular point. It is a highly expert matter to decide whether knowledge has developed to the stage at which the deployment of massive resources at this point will remedy the deficiency,[1] or whether it is still so indefinite that research can only

[1] Looked at scientifically, the development of the atomic bomb in the short space of five years was a remarkable achievement. Primarily, however, this depended on the correct assessment at the outset that the relevant scientific knowledge had developed to the stage at which there could be

continue to probe the possibilities. Closely related to this is the identification of opportunity. Such is the present degree of specialization that it is not possible to assume that research workers will always be aware of the full implications of their findings. Only if close scrutiny is maintained of developments in the province as a whole can one guard against the possibility that significant opportunities will be passed over and the need for reinforcement overlooked. Conversely, it is often only in the context of the whole province, or at least sequence, of knowledge that false trails can be detected. Only too frequently projects in particular subjects started with high hopes become increasingly disappointing, or once fashionable subjects are revealed as becoming intellectually unproductive. Those engaged are rarely the first to realize this. Lastly there is the brute factor of available resources. Today, the costs of scientific research are rising to such levels that choices are becoming inescapable. If these are not made on the basis of informed assessments of the relative importance of different developments to the progress of knowledge, then they will undoubtedly be made on other grounds, and, public expectations from scientific research being so high, these are likely to be influenced more by wishful thinking than by an objective scientific appraisal of the situation.

It would seem, therefore, that if scientists are to retain control of professional aspects of scientific development, and thereby be in a position to discharge their particular responsibilities to the society of which they are a part, they will need increasingly to take conscious and deliberate thought for its future. They will need to think increasingly in terms of the intellectual importance of the objectives of particular research projects and not merely in the traditional terms of intrinsic interest. They will have to ask such questions as what would be the significance for the development of scientific knowledge if the research proposal in question

---

reasonable confidence that, given adequate resources, the development of controllable energy from atomic sources was feasible.

The development of an effective vaccine against poliomyelitis was more problematical. This took much longer and it was only achieved by deploying heavy resources widely and upgrading the field of virology in general.

THE ORGANIZATION OF SCIENTIFIC DEVELOPMENT

were to succeed. These questions all involve discrimination, and the ability to discriminate depends upon the existence of criteria external to the particular matter under consideration. In only a limited and technical sense, therefore, can the significance of any research project be judged in relation to the discipline with which it is identified. Its larger significance derives from different considerations, namely the importance of its potential contribution to the intellectual continuum of which it is a part, and which itself is the expression in thought not of a subject but of a province of natural knowledge. If, therefore, policy for the development of scientific knowledge is feasible, it is along these lines that we should seek it.

But to what extent can we feel justified in going in this direction at the present time? Any person with experience of the way that knowledge develops is only too well aware of the limitations of his foresight. At any time developments may occur that alter the whole outlook and disclose intellectual needs that previously had seemed of little moment. Yet, if we are not to be entirely inhibited from any attempt to cope with the situation, we cannot leave matters in this state. Nor in the present state of scientific knowledge in most provinces would we be justified in doing so. Fortunately the solution is ready to hand.

Over a country as a whole there are many sources of individual initiative and a wide diversity of research interests. At any particular stage in the development of scientific knowledge, not all are in the forefront of attention. It is therefore of the first importance that an adequate range and diversity of scientific interest be kept in being, and it is a prime responsibility of any central research organization to see that this is so. Without the insurance given by such a background, no organization could responsibly take the risk of backing its judgment on the emphasis to be given to the support of the different subjects within its province. Given such a background, however, it is now both allowable and possible to promote a purposive policy for development. But policy must cover the whole of the particular province. It cannot afford to be biased either towards the specialized periphery or towards the unspecialized centre. Differences in intellectual emphasis within a province require justification, not in principle and once-for-all, but in the context of an over-all view and the changing contribution of the component subjects of a province to the continuum of knowledge which this expresses.

As I see it, therefore, scientific knowledge has now developed to the stage where purposive policy for its further development is becoming feasible, and provided an adequate range and diversity of research interest is kept in being as a background to this, it is justifiable to take positive steps towards its implementation. First and last, this is a professional problem, for it is only by comprehending the intellectual interrelationships of the component subjects and the implications of the developments within each, that the necessary synoptic view can be attained. This is a matter that calls for the highest scientific judgment and knowledge. Only if policy for the development of any particular province of scientific knowledge is under the control of a body the members of which are predominantly themselves research workers of distinction and which also has the necessary resources and independence to give effect to its conclusions, is research policy in any effective or safe sense possible.

III

It is evident, as we have previously remarked, that today it is impossible to think effectively of the requirements for the development of any province of natural knowledge on a scale less than that of the total resources of a country available for these purposes.[1] Yet in no province are the resources for research itself concentrated in any single agency. In all we find research in association with universities, research in association with practical activities and research directly related to the central organization. Each of these agencies differs from the others in the essential purpose for which it exists and consequently in its attitude to scientific development. Further, in the first two types,

[1] In the case of some problems even this is not enough and we must think on a global scale.

research comes from a multiplicity of independent sources each of which is entirely at liberty to determine the nature and direction of that which it undertakes. If, therefore, there is to be any unification of purpose behind the development of research in any province, it is only to the central organization with its country-wide responsibilities that we can look. The relationship between a central research organization and the other agencies that undertake research is, in consequence, a matter that at the operational level is of prime importance.

In general, central research organizations deploy their resources in two broad ways. The first is by means of relatively small short-term grants to assist individuals in other agencies to develop their personal interests. It is on these that reliance largely rests for the promotion of that range and diversity of research that is the necessary safeguard to any positive policy for research. The second is by more massive long-term support either to work in their own institutions or to that in other agencies; and it is to massive support of this kind that they must primarily look for the implementation of any positive policy. Thus, in pursuit of its primary responsibility for the comprehensive and balanced development of knowledge in its particular province, the operational role of a central research organization is essentially complementary to that of the other two broad types of organizations undertaking research. It will, in consequence, seek to act through these whenever it can. An assessment of the extent of the potentialities in each lies, therefore, at the basis of any consideration of their relationship.

Today, the universities of most countries are the foundation upon which their provision for research is built. Because of their responsibility for higher education, it is from them that new talent emerges. By their tradition of individual autonomy in intellectual matters they are the source of a wide range and diversity of scientific research. And because of their multidisciplinary structure they provide an environment conducive to the stimulation and development of original thought. In respect of resources for the support of research, the universities, *qua* universities, because of their concern with higher education, have funds at their own disposal. In this way, members of their staff are assured, as of right, of a basic level of support for their personal research interest. Beyond that, the understanding is that support shall be sought from outside organizations with an interest in the research in question. Collectively, therefore, the universities provide the major opportunity to ensure that wide background of research against which to implement a positive research policy. Individually they offer one kind of environment in which massive support can be deployed when policy requires the more intensive cultivation of particular parts of a province of knowledge.

At the other extreme, where research abuts on advanced practice, there is the industrial corporation. The primary purpose of research under their auspices is necessarily directed to further the commercial interest with which they are concerned. As autonomous and independent bodies the nature and extent of the research they undertake depends entirely on their own initiative and judgment, and, although enlightened self-interest may lead them to penetrate considerably towards unspecialized research, their ultimate purpose remains. Analogous considerations of external interest other than the advancement of knowledge apply to all research arising out of practice, whether this be in a profession, industry, administrative department or in the public or military services. Nevertheless, we cannot afford to neglect the significance of work at this level. As we have stressed repeatedly, perhaps the major weakness in thinking about scientific development at this present time is the extent to which the significance of contributions to original knowledge from these sources is discounted. The fostering of such research and the integration of its knowledge with that of work at more unspecialized levels must consequently be a major concern to any central research organization. But just as the universities should have resources at their own disposal so, at the level of practice, local support should be assured to the extent that initiation can be taken. Clearly the way of doing this will vary with the nature of the practical activity. Given this, however, a central research organization can use its own resources to complement effort either locally or centrally in support of any policy it has for the balanced development of its particular province of knowledge. But a central research organization cannot expect to implement its whole policy entirely through the agency of others. Massive research projects can only be put down with the agreement of the host institution and this will naturally be dependent on the project being compatible, both in scale and staffing, with the purposes for which the institution in question primarily exists. Further, the kind of research proposed depends upon local interest, individual in the case of the universities, institutional in the case of industry. If, therefore, the purpose of a central organization is not to go by default, it cannot rely entirely on the chance that the spontaneous interests of others will meet its needs. It must ensure that it is in a position to take the initiative whether this is to deal with gaps or weaknesses in effort that are retarding development or to exploit opportunities that it has identified. For these purposes it will require staff of its own. In general, it will need such staff to develop subjects that are being neglected or inadequately developed by others, to support new subjects that have not yet established their claim to acceptance, to undertake research on questions of public interest which are either outside the concern of other agencies or require developing on a scale beyond their ability, to provide for outstanding men who do not fit readily into existing niches and to reinforce endeavours that merit developing more intensively. Many projects of these kinds are, by the prevailing standards, either unorthodox or unfashionable. If men are to be attracted to undertake them then necessarily they must be shown a career. To do this is a primary duty of any central research organization. Nevertheless, such organizations would be well advised to ensure that careers with them are interchangeable with those of other agencies to which they are related so that a man may follow any natural change in his interests.[1]

---

[1] Young men in the first flush of their thought usually ask for nothing better than a whole-time research career. By their middle thirties, however, a substantial proportion are coming to entertain doubts if this is their true métier. Most of these wish to continue to do some research, but preferably

In some countries the staffs of central research organizations work entirely within its institutes. In others, like ours, they work partly in institutes and partly in teams placed in or alongside the other agencies for research. Research institutes, if they are large and include an adequate spread of disciplines, can be highly effective for their purpose. By their freedom from responsibilities other than research they can organize to this end rather than in conformity with some other purpose. In consequence, they can readily develop new groupings of interests. Further, if they are not organized on a basis of established departments, they can both promote and sustain that interdisciplinary collaboration which is so necessary to the intensive attack on major problems.

The second device, that of placing teams of staff in other agencies, was pioneered by this country. It has proved of mutual benefit. Such teams, or research units, may be under the direction of a member of the central research organization's staff or be attached to a member of the staff of the agency in question who then serves in an honorary capacity. Provided that such research units are built round an able man and not continued beyond either his tenure or the purpose for which they were formed, they constitute an effective and deployable instrument of policy.[1]

It is by operational measures along these broad lines that a central research organization can realize its double purpose of ensuring that an adequate range and diversity of research is in being over the country as a whole and that, at the same time, a more purposive policy can be pursued in regard to the developing opportunities and needs of research. Further, by the same means it equally ensures that it maintains a living and operative contact

---

not whole-time. It is in the interests of all concerned that alternatives should be readily open to them.

[1] Naturally the device of attaching departments of the central research organization to a professorial member of a university staff has considerable appeal. By it the professor who serves as its honorary director is assured of a continuing private research income and staff insulated from the vicissitudes of local considerations, and the university of a major department at little or no extra cost to itself. On occasion when policy changes this may clearly give rise to difficulty but, nevertheless, the efficacy of the arrangement outweighs this.

with the whole range of experience from the most specialized to the unspecialized in the province of knowledge for which it is responsible and, thereby, is in a position to take a synoptic view of the whole.

IV

There is, however, another highly important instrument for the promotion of scientific research requiring consideration. That is the scientific society. These are self-governing organizations set up at the instance of research workers themselves. Broadly speaking, they are of three kinds: the specialized society with interests in a particular subject; the professional society or college; and the scientific academy.

The specialized scientific society exists primarily as a forum wherein workers with interests in the subject in question can meet to communicate experience in an atmosphere of healthy criticism. In a very real sense their contribution is complementary to that of a central research organization, for, whilst this seeks to integrate knowledge so to speak 'vertically' throughout its province, they seek to integrate thought at the subject level and particularly in respect of those subjects that run horizontally, like strata, across many sequences and even several provinces of knowledge. Such societies, by their discussions and journals, have in consequence played an indispensable role in the communication of knowledge and in the formulation of theoretical understanding, not only within their own community but internationally through their association with similar societies in other countries. They are, and will continue to be, an essential instrument of scientific development at the subject level.

The primary purpose of the professional society, association or college, is the organization of its profession. Of necessity it is concerned with the developing knowledge in its particular field both in respect of its responsibility for certifying the competence of its members and its responsibility for promoting the standards of professional performance. To the findings of research, it must always be alert. Research, however, is not its primary purpose and, although it may promote this to some extent, this can only be as an incidental to its other activities.

Scientific academies have played an historic role in the development of scientific knowledge. At first when scientific knowledge was less they were, in effect, specialized scientific societies taking in all natural experience. With the growth of knowledge, however, this role has largely been taken over by the multiplicity of specialized societies that has inevitably come into being. In this respect their role is now that of directing preferential attention to developing fields that they have identified as particularly significant. But in the course of time they have developed two further functions. They have come to be regarded as the acknowledged guardians of the standards of excellence in scientific knowledge and thereby set levels of achievement and integrity that largely account for the high respect in which the scientific community is now held. And they have come, as a result of the scrupulous objectivity and deep knowledge manifested in their opinions, to be accorded a high degree of authority, not only nationally but internationally, in all matters of scientific knowledge as such. In relation to the future, these two functions are bound to develop. As scientific knowledge becomes more important to the community, so do the needs for high standards in its pursuit and for independent opinion on its development.[1]

It would be incomplete, however, if these considerations were left without reference being made to the important role that, in western countries, foundations dispensing private funds have played in this context. Not only have they given support to scientific academies and professional colleges but many have themselves promoted studies or research to develop knowledge in

---

[1] In some countries, notably the U.S.S.R., the functions of central research organizations and those of a scientific academy are, in effect, brought together in one organization. The advisability of bringing so much authority together into one focus is debatable. If a country has considerable leeway to make up there may be much to be said for this. If, however, a country is already at the frontiers of scientific development there is equally much to be said for ensuring the existence of an independent authoritative source of criticism distinct from the executive organizations for research in the form of a scientific academy.

those fields that were of special interest to them. By so doing they have been a most valuable adjunct.

V

Having now given our reasons for believing that, within limits, research policy for the development of scientific knowledge is feasible, and having indicated the means by which such policy could be put into effect, we can now approach the second problem; that of relating such knowledge to the needs of human societies. Clearly the organizations concerned with scientific development will have a large part to play in this respect. The first question to be answered is, therefore, what responsibilities has a central research organization to the society of which it is a part over and above those for developing scientific knowledge in the province to which it is related?

To answer the question in the most general terms, one could say that such an organization is concerned with assessing the significance of developing knowledge in its province for the social purposes that are related to this. To answer the question more precisely, however, we need to look at what this implies in practice. I believe that it entails three things: the identification of the implications of developing scientific knowledge; the assessment of the feasibility of meeting a particular human need by scientific means; the objective appraisal of the consequences of adopting a particular course of action involving scientific considerations. These are each and all purely professional matters. As such, only scientists can make them.

The full implications of a new development are not necessarily immediately obvious. It is a commonplace to lament the long gap that occurs between discovery and application. To a very considerable extent this is due to the limitations of awareness consequent on the high degree of specialization now forced upon men at all levels from the most practical to the most recondite. It is a particular function of a central research organization that, spanning as it does all levels in respect of a particular province, it should seek to promote awareness throughout the whole range of its knowledge. On many occasions such implications reveal new opportunities; on others, mistakes either intellectual or practical.

The obverse of the problem of identifying opportunity is that of feasibility. Now, when scientific progress has aroused such expectations of further development, desire frequently threatens to outrun the possibilities of performance. The existence of a human need is no assurance that it can be met by scientific research. Frequently, dispassionate assessment reveals that, in the existing state of knowledge, hopes to this effect are groundless. Only, however, if such assessments have been made by those who are both professionally competent to do so and known to be independent of other considerations, are they likely in present circumstances to be publicly accepted.

Assessment of the consequences of practical policies involving scientific considerations arouse even more acutely questions of public confidence. In the middle 1950s, for example, the hazards to human populations arising from the exploding of atomic bombs became of acute concern. Not only had this implications for the individual but it had also political implications of serious import both at the national and the international levels. In the existing state of ignorance, anxiety ranged wildly and political reassurances were of no avail. Only when scientists themselves produced a demonstrably independent assessment of the consequences of such explosions was confidence restored. This was a major example of the problem. Answers to lesser problems of more limited significance are increasingly required in the scientific age into which we are now entering. All such assessments include in their scope ranges of scientific knowledge from the most specialized to the most unspecialized. Attention confined to either extreme of this range is insufficient. Only by objectively integrating the whole can a result that might be dangerously misleading be avoided.

These, then, the assessment of scientific opportunity, feasibility and consequence, are responsibilities that scientists, and only scientists, can discharge. Further, only rarely, in these days of specialization, are they within the competence of a single individual. For their full appraisal, the resources of an organization including the whole necessary range of individuals with overlapping interests, is more often required.

But this is only one side of the problem. To be fully relevant such assessments must be informed not only of actual or contemplated practical activities but of the way in which the machinery of society operates. In a very real sense those who would seek to integrate scientific knowledge into social activity must learn to speak not only the language of science, but also that of administration. Only if their awareness extends beyond their particular professional interests, can they hope to have that understanding which forms the basis of any integration. It is essential, therefore, that any organization that aims at integrating any province of scientific knowledge into social thinking should have, at the extreme of its specialized activities, the closest collaboration and actual working contacts with the social organizations, be these industrial, professional or governmental, to which its particular knowledge is relevant. Given this, and given independence so that it commands the confidence of public and scientists alike, such an organization can fill the need that the rapid development of scientific knowledge has brought into being. This was the new need that was so clearly foreseen by Haldane, Addison and Morant, and it was to meet this that they devised the scheme of independent central research organizations with executive authority and under scientific control as a complement to the other agencies both scientific and administrative that already existed.

VI

At the outset, I gave as my justification for writing this essay the belief that what we call 'science' had reached a crisis in its affairs, and that unless scientists took conscious and objective thought for its future, its further development might well be in jeopardy. I am by no means alone in this belief, but different people see the situation in different ways.

Essentially, as I see it, the situation that now faces us has been created by the very success of scientific research itself. It has now been demonstrated beyond doubt that, by means of such research, man can expect increasingly to extend his control over natural phenomena. As a result, scientific research has ceased to be a purely personal interest and become a vital matter of public concern. It is this, not the demands that are made on its behalf, that accounts for the escalation in the support it has received during recent years. To close one's eyes to the hard fact and to maintain that all would continue to be well if research were adequately supported and left to its own devices is not only unrealistic, but dangerous. Scientific knowledge is now too important to society for scientific research to be left outside public concern in this way. Either scientists themselves must show that they are capable of controlling its development or all human experience shows that it will be done for them.

And there is a further and perhaps more important consideration. Scientific research took its start in the endeavours of men to understand and control their natural environment. It owes its support and privileged position in society, like the traditional professions, to the expectations that it will contribute to the common benefit. In a sense, therefore, research workers, like members of these professions, have by adopting research as their profession, incurred moral as well as intellectual responsibilities. Intent that went no further than self-interest may have been excusable in the past when scientific knowledge was so rudimentary that purpose could be little more than wishful thinking. Now, however, the scientific movement is becoming increasingly effective and scientists, with their special knowledge of the potentialities, are in consequence incurring increasing responsibility for the direction in which it develops. Although, therefore, it is an essential condition of effective creative work that in pursuit of a particular project it should follow its own inspiration, that does not relieve the creative worker of responsibility for the intent with which he undertook it or for assisting the development of which it is a part. Even if enlightened self-interest did not require that scientists should take into account the public expectations that scientific development has aroused, these obligations would still, in principle, be there and constitute a claim on the services of scientists to give to the development of the particular over-all purpose to which their interests are relevant the skilled guidance that they are now in a position to give.

At the present time the most urgent problem facing the scientific community is that of so organizing research as to take full advantage of the new feasibility of achievement that its past success has shown to be increasingly possible. Essentially this requires that the component subjects of knowledge be developed not only for their own sake, but also with forethought for their relationship with each other. This is a task calling for the best scientific skill available. The traditional instruments for its realization are research in association with universities and research in association with practice. Each of these is distributed among a multiplicity of autonomous agencies, all of which have their individual interests, loyalties and policies. These are indispensable assets, but, by themselves, they are now not enough. Nor, when scientific research has become of such moment to society, could their essential autonomy survive the demands that would be made upon them were they to stand alone. To complement their essential contributions a new type of organization has now come into being. Intellectually this corresponds to a whole province of natural knowledge, and being nation-wide in its scope it is in a position to take the synoptic view required. It thus offers to the scientific community the kind of additional instrument now needed for the further development of scientific knowledge and for the discharge of its social function. Its proper development is consequently among the major responsibilities falling to scientists.

As I see it, therefore, in the situation into which we, as scientists, have now entered, the conceptual basis of units of organization for the further development of scientific knowledge is the province of natural knowledge, and the operational instrument corresponding to this is a central research organization with nation-wide responsibilities. In the present circumstances the provinces of knowledge that require representation in this way are, I have suggested, biomedicine, bioagriculture, materials, energy, transportation, communications, structures, natural environment and social sciences. These are the units, both intellectual and instrumental, from which organization for scientific development can be built. Policy in respect to the development of knowledge in each of these provinces is entirely a matter for professional scientific judgment, and the responsibility for this can only be vested in an essentially professional body.

It falls now to consider the means of organizing to this end. Before doing so, however, it is necessary to draw attention to a factor that, from now on, will come increasingly into consideration.

Hitherto we have been predominantly concerned with the requirements for scientific development arising out of the nature of scientific knowledge itself. But in approaching questions of organization, we have also to take into account the particular social structure through which this has to express itself. This differs widely from country to country so that the details of organization will differ also. Nevertheless, natural phenomena are the same everywhere and the basic requirements to be met are universal. Although, therefore, the limitations of my own experience will impose upon me the need to discuss these matters to a considerable extent in terms of the social organization with which I am familiar, it is hoped that this will not obscure the general principles involved and that others will readily translate these into terms of the social structure with which they are conversant.

✴ II ✴

ORGANIZATION AND RESEARCH POLICY

I

THE ROLE OF A CENTRAL RESEARCH ORGANIZATION that has emerged from the preceding considerations is a dual one. It is that of a body that is related on the one hand to the development of scientific knowledge and, on the other, to the needs of society for such knowledge. If, therefore, such a body is to fulfil its function effectively, these dual requirements must be reflected in its constitution and in its placement in the machinery of government. These problems can most easily be approached by considering first the constitution of the governing council appropriate to a body with these responsibilities. We can thereafter consider the form of its executive machinery, the relation of such organizations to each other and finally their integration into the structure of government.

II

Broadly speaking there are two types of council or committee, each of which can be distinguished from the other according to its purpose. The purpose of the one is to reconcile conflicting interests, and for this the principle of representation of interested parties is essential. The other is the expert body, the purpose of which is to promote knowledge. On this representation of interest, be this institutional or political, has no place. The only criterion of eligibility for membership on councils of experts, is personal merit in respect of a relevant aspect of knowledge. If, therefore, the Research Councils governing the different central research organizations are to command the confidence either of government or of scientists, they must be composed predominantly of men who by their own contributions to the development of scientific knowledge, have won the respect of the relevant scientific community. Collectively these should cover the range of disciplines that span the particular province of knowledge. Further, appointments should be made only for limited periods, with a prohibition against immediate eligibility for reappointment, so that the membership of the particular council may continue to reflect the changing needs of developing knowledge in the province for which it is responsible.

Apart from the members who are there because of their scientific attainments, a minority of members should be 'laymen'. These should be men experienced in the handling of large affairs and who have shown by their lives their ability to live with the consequences of their own decisions. The purpose of such membership is twofold. The first and obvious reason is to command the confidence of the non-scientific public. The second is a matter of practical working. These councils are primarily composed of experts, and experts in one subject are always hesitant about questioning the opinion of an expert in another. An experienced layman can readily ask for elucidation, often to the manifest relief of his expert colleagues and, even more important, to the restoration of perspective.

But, if the research council governing a research organization is to command the necessary degree of confidence in its objectivity, considerations of interest, political or otherwise, can play no part in appointment to its membership. On the other hand it is only right that, in respect of a body with such public responsibilities, the powers of veto and dismissal should rest with the government that provides its resources. But the choice of experts is itself a most expert matter. Now that scientific considerations have become so complex, only those who have close and professional appreciation of men in the relevant fields are likely to be able to advise on this. In regard to scientific members, therefore, consultation with the research organization itself and the national scientific academy should be obligatory.

No body like the governing council of a central research organization can, however, function without a secretariat. On this they must rely for the implementation of their policy and all the details of executive action that go to build this up into achievement. This requires a range of knowledge and skill in its own right. But it is not beyond the capability of a scientist to acquire this. In the secretariat the key role is played by the chief executive officer. Traditionally in this country (although not in many others) the expert has been excluded from taking executive policy decisions and his role confined to that of giving advice to professional administrators. Today when scientific knowledge has become so complex that its appreciation may tax even the most expert, it might be thought that the day of the traditional attitude to matters in which such expert knowledge is involved, had passed. In respect of scientific research this country has, however, been fortunate, and for this they have largely to thank Haldane and Addison. From the outset it has been accepted that the chief executive of its research councils should be a scientist of distinction who combined in one person the roles of administrator and chief professional adviser. Nevertheless, it is doubtful if the scientific community have as yet fully appreciated the extent to which scientific development depends, and will increasingly depend, upon the willingness of suitable scientists to devote themselves to such tasks or the conditions under which such men should operate.[1]

Until recently in this country, the chief executive of a research organization occupied the post of Secretary to its governing council. More recently the view has been put forward (and in some cases implemented) that he, rather than one of the council members, should be chairman of the council.[2] In retrospect, I am doubtful about this suggestion. Committed as he must inevitably become to the implementation of particular policies, there is much to be said for having another person as chairman of the policy-forming body. Further, any particular man has necessarily a particular viewpoint, and a chief executive who was also

[1] P. Kapitsa: 'The Future Problems of Science' in *The Science of Science*, p. 126. Penguin Books, 1966.
[2] *Report of Committee of Enquiry into the Organization of Civil Science*, 1963, para. 110.

chairman would, by reason of his longer tenure, tend to offset the periodic refreshment of policy that should come from the changing membership of the council. Looking to the future, therefore, my present view is in favour of the chairman being selected from among the council members and the secretariat of the council being modelled on that of the special organizations of the United Nations. The chief executive, although a member of council *ex officio*, would then be in a position corresponding to that of a director-general and, like him, appointed for only limited periods of time, preferably by the council itself.

In general terms, therefore, it is on such broad lines as these that one envisages the internal structure of a central research organization. These are, I believe, consistent with such organizations filling the role not only of scientific bodies but also of autonomous executive authorities.

III

At the level of actual research policy, therefore, the situation is that we in this country have evolved a series of autonomous research organizations. These, at their inception, had relatively few interests in common. With the growth of scientific knowledge, however, the position has changed. The question of the relation of these organizations to each other consequently now requires consideration.

Broadly speaking, the problems common to the several research councils are of two kinds: those relating to scientific collaboration between organizations with adjoining interests; those relating to common problems of management. The following examples will serve to illustrate the nature of these and to indicate the further measures that are now necessary.

Man is a gregarious animal and many of his ills derive from the strains of living in human societies. Collaboration between the central research organizations concerned with biomedicine and social sciences respectively is essential for neither organization alone has either the knowledge or the expertise to cover these junctional problems. In recent years we have entered on the age of nuclear power with its potential hazards to the health of present and future generations. Only by the collaboration of medical men, biologists and nuclear physicists can these be faced. Space exploration has raised new problems in communications, materials and engineering that call for co-operation between the central research organizations primarily responsible for each of these. Transport at increasingly high speeds is a feature of modern societies, and the control by the human operator of machines designed to this end is an essential consideration in its promotion. The joint resources of the research organizations concerned with transportation and biomedicine are required for such problems. These examples could be multiplied many times. Situations ranging from the transient to the continuing are common. Provided that the different central research organizations have been properly identified with particular provinces of knowledge, effective scientific collaboration and any necessary joint enterprises should give rise to little difficulty.

The problem of collaboration in respect of those common interests that at the subject level cross several provinces of knowledge is already largely provided for. The specialized scientific societies that have come into existence for the very purpose of promoting communication and the development of understanding at these unspecialized levels have already ensured the essential basis for this. Drawing their membership as they do from those in any province of knowledge to which the subject in question is relevant, the necessary intellectual cross-fertilization is assured by the spontaneous interests of those concerned. Given that the particular subjects find expression in the composition of the governing councils of the relevant central research organizations, their integration into scientific development can readily be effected.

In relation to a structure of organization that depends upon a confederation of several autonomous agencies, the question of inadvertent deficiencies in the over-all coverage quite properly arises. At first sight this might seem to be a very real danger in respect of scientific development, and it has been a matter for surprise, or even on occasion suspicion, when enquiry has failed to reveal gaps in coverage of any moment. But there are two reasons for this. Central research organizations never lack for gratuitous advice either from individuals or scientific associations, and it is in consequence unlikely that any genuine deficiency of coverage, either within or between them, could escape being brought forcibly to their attention. The second reason is that the whole trend of scientific development is towards increasing coalescence. Individual subjects may still be far apart, but, when knowledge has reached the stage when it is intellectually feasible to effect a junction, the spontaneous interests of research workers on either side of the gap will ensure that sooner rather than later it is closed.

A greater danger is in a sense the opposite of this. It is that with the growth of knowledge at all levels, a central organization may come to have too wide a remit. A typical example of this was the situation in the Department of Scientific and Industrial Research, which at the time of its dissolution had become responsible for several provinces of knowledge rather than one. A very similar problem may quite probably arise in the future in respect of the central research organization concerned with the social sciences. When such a situation does arise, however, the solution is to split the province in question so to speak vertically along the natural lines of cleavage apparent at its periphery and to form new research organizations corresponding to each of the adjoining provinces of knowledge thus differentiated. It is never to cut the province horizontally and so divorce the unspecialized from the specialized knowledge on which it depends not only for its *raison d'être* but also for its continued awareness of unedited natural experience.

The preceding are all considerations arising out of the nature of scientific knowledge itself. But there is another category of problem, that concerned with actual management. In the matter of their operating, none of the central reseeearch organizations is self-sufficient. All act to a greater or less extent through the agency of universities; all give grants, all support research training, all employ scientific staff, and all have relations with scientifically based practice be this professional or industrial. Common problems abound, and clearly there should be some consistency in the ways with which these are dealt. Appropriate inter-organization committees, staffed by those who have first-hand knowledge of the problem in question, can readily deal with matters internal to the system of research councils. Bodies including representatives of other interests can similarly deal with matters involving external relationships.

It is evident, therefore, that at the organizational level of central research organizations there is a need to promote collaboration and to apportion responsibility for research in the subjects of common interest, within the general perspective of accepted national purpose. In the past, when such organizations were few and their operations relatively simple, the frequent and informal contacts between the scientists who were their chief executive officers largely sufficed to ensure this. Now, however, when central research organizations have increased both in number and in the scale and complexity of their operations more explicit measures have become necessary. Accordingly a Committee of the Chief Scientific Executives of the several central research organizations, meeting under the chairmanship of one of their number, should formally be set up to promote collaboration and operational policy and to give collective advice to Government at the executive level. Such a chiefs-of-staff committee is now essential as has similarly been discovered in the universities where, in this country, it takes the form of the Committee of Vice-Chancellors and Principals.

IV

We can turn now to the question of the relation of central research organizations to Government.

From the point of view of Government, scientific knowledge is important in so far as it bears upon national policies. In this connexion it is important to distinguish between departmental policy and Government policy; between policy that concerns the special responsibilities of a particular department of Government and falls within the authority of that department to decide, and policy that involves Government as a whole and which requires for its implementation a collective or Cabinet decision. Some examples from the medical field of research in relation to departmental policy and then of research in relation to Governmental will make this important distinction clear.

A health department[1] may wish to embark upon a national policy of preventive vaccination to protect children against a common infective illness. The devising of vaccines, questions of their efficacy, safety, dosage and the timing of their administration are all matters that depend on the findings of biomedical research. Transplanting of organs is a problem exciting great public interest and hope at this present time. A National Health Service must have a policy in regard to the provision it should make for this. But knowledge of this subject is only in the process of evolution. Departmental policy can go only so far as the results of research justify; and assessment of the significance of research in this complex subject involves a whole range of biomedical disciplines from the most specialized to the most recondite. Again a department responsible for the conditions under which labour operates may have become concerned about the safety of an industrial process involving the use of certain chemicals. The decision whether or not to forbid the process and, if not, what precautions to take, will turn primarily on the findings of research in respect of the toxic action of the chemical in question. The development of atomic energy as a source of power raised important questions about the protection of workers exposed to ionizing radiations. To answer these, a whole range of biomedical research in depth was required. Its results were of profound significance to the department concerned, not only from the point of view of health, but also, because of the consequent requirements for the design of atomic plants, to considerations of the economics and engineering of this form of power. A department concerned with transport or aviation must necessarily pay close attention to the factors that facilitate and those that impair the skill of the human beings who

[1] In the United Kingdom there are separate health departments for England and Wales, Scotland and Northern Ireland rather than one single Ministry.

operate the potentially dangerous machines with which they are concerned. The feasibility of military operations in tropical countries may well turn on means being found to protect against endemic disease or the effects of climate. Biomedical research in regard to the effects of noise on health is of concern to no less than seven departments of government, excluding those of defence.[1]

All these are instances of situations in which decisions on policy in any one of these very different departments may turn on the findings of biomedical research. From the point of view of such departmental policy, the findings that are of interest are the end results of research that relate directly to the particular practical situation. But these results rest on the promotion of a whole range of research activities from the specialized periphery to the more unspecialized regions of the biomedical province. None could have been obtained save on this basis. The province of knowledge that we call biomedical is central to all such particular concerns. This is what Haldane had in mind when he distinguished research for general use from the particular operational research of individual departments and this is what led Addison to centralize biomedical research proper so that it could be available to all rather than segregated in a particular user department.[2] Organization at this level should, therefore, present little problem for it is no more than a matter of ordinary day-to-day working relationships between government departments and the different central research organizations.

[1] T. Dalyell: *New Scientist*, 1969, no. 42, p. 196, last para.

[2] 'But it is certain that if the [Medical Research] Committee were known to be working in specially close relation to a progressive Ministry of Health, and also to be substantially under the control of the Minister, all other Departments would begin to object to using the Committee and giving it full information, and would do their best to conduct the whole of the medical research which they required through their own officers. This would prevent any single body, such as the Medical Research Committee, from having under view the whole of the medical research which was being done on behalf of Government, and would make impossible the proper distribution of the work between the separate Departments and the medical research body which should be the common helper of all Departments.' Memorandum on the Ministry of Health Bill, 1919. Work of the Medical Research Committee. Cd 69. H.M.S.O., 1919, paras. 14 and 15.

ORGANIZATION AND RESEARCH POLICY

Leaving now the level of specialized departmental policy, we can look at three examples of situations in which the findings of biomedical research bore directly on policy of Government itself.

Reference has already been made to the situation that arose when it was discovered that radiostrontium, derived from the 'fall-out' of exploding atomic weapons, was accumulating in human bones and particularly in those of young children. Understandably this raised acute public alarm and, for Government, major questions in regard to foreign and defence policy. In this country the Prime Minister requested, as a matter of urgency, that the Medical Research Council should produce a comprehensive assessment of the scientific aspects of the situation. A glance at their report will show the range and depth of knowledge required, and the research that had to be undertaken to this end.[1] Again, because of our island situation, the rationing of food became a major consideration in two World Wars. An effective policy to deal with the successively anticipated stringencies could clearly only be worked out on the basis of scientific knowledge of nutritional requirements. Contingent on the acceptance of this agricultural, shipping and defence policies were all influenced to a greater or less extent. Lastly, in the early 1950s, the first evidence of the association between cigarette smoking and the development of bronchial cancer was produced. The tax on tobacco in most countries being a major source of revenue, these findings inevitably raised serious fiscal questions.

I have drawn these examples from the medical field because of my close personal knowledge of the nature of the issues involved both scientifically and politically. Doubtless others with comparable experience in other fields could match these or show where lack or unawareness of scientific information had handicapped Government in taking effective policy decisions. I hope, however, that I have said enough to make the essential point. Under modern conditions, scientific knowledge has become so important in its implications for numerous general aspects of Government policy that, if informed decisions are to be taken,

[1] *The Hazards to Man of Nuclear and Allied Radiation.* Cmd 9780. H.M.S.O., London, 1956.

the organizations for scientific research must be related directly to the centre of Government. Only if such a direct relationship with independent organizations of this kind is available can Government in the collective sense feel assured that it has access to the uncommitted advice, assessments and interpretations it requires for its purposes or that discoveries in research with implications for its policies will be speedily identified and brought to its notice.

In the next chapter I shall give the reasons that have led me to the view that, in the context of this country, there should be a Committee of Cabinet made up of Ministers whose departmental responsibilities involve substantial scientific considerations, and that this should be under the chairmanship of a senior Minister of Cabinet rank without departmental responsibilities. It is to this Minister that the central research organizations should be responsible and it is through him that such organizations should relate to the centre of Government in respect of the political implications of developments in the particular provinces of knowledge for which they are respectively responsible.

This then is how I conceive the place of a central research organization in the structure of the machinery of government on the one hand and in its relation to the scientific community on the other. It remains now to consider the over-all problem of national policies in respect to the balance of scientific development.

* 12 *

NATIONAL POLICY AND SCIENTIFIC DEVELOPMENT

I

NOW THAT ALL DEVELOPED COUNTRIES recognize the importance of scientific knowledge for their future progress and when scientific research has become primarily dependent on national funds for its support, national policies for scientific development have become of crucial significance. It is this that accounts for the attention now being devoted both by governments and by scientists to what is generally called 'scientific policy'. It is essential, therefore, that we should be clear as to the nature of the problem at the national level.

The exact scope of the meaning attached to the term scientific policy varies considerably. In general, however, whatever the scope, the concept rests on the tacit assumption that it is possible, on the basis of scientific considerations, to arrive at a policy for the development of scientific knowledge in all its aspects which would enable its requirements to be specified and decisions reached on the extent to which its different parts should be supported. It is hardly necessary to stress the importance of this concept or its significance for scientific development. If it is indeed possible to work out a general scientific policy in this sense then the parameters within which national policy must operate will be clear, and the difficulties of organizing to this end largely resolved. If, on the other hand, no such policy is in fact feasible then endeavours on this basis could lead us seriously astray. Examination of this problem is, therefore, an important preliminary to any consideration of policy at the national level.

II

Reasons have already been given for accepting that, within limits, it is now feasible to formulate policy for research within a particular province of scientific knowledge. The foundation of this view is that, within any particular province, the component subjects stand in a coherent relation to each other so that the province as a whole forms a logical intellectual continuum. It is thus possible to relate the contributions of one subject to another, on the basis of genuinely scientific considerations, and thereby to arrive at a policy for the balanced development of knowledge in the province as a whole. The question now before us is, can we go beyond this and produce a policy that is equally scientifically based for scientific knowledge in general?

If one looks at the content of the knowledge derived by research from the different provinces of natural experience one is driven inescapably to the conclusion that this is as diverse as the categories of natural phenomena that have been studied. It is this I believe that Karl Pearson had in mind when he said that the unity of science (like logic) lay in its method, not in the materials to which it was applied. One has only to consider the possibility of comparing say physiology with nuclear physics or the results of cancer research with those of space research to see the force of his contention. Scientifically speaking these subjects are incommensurable. But if there is no scientific basis for relating one category of scientific subjects to another there can equally be no scientific basis on which to construct an embracing policy for their development.

The inference from this conclusion is inescapable. It is that scientific policy in any meaningful sense of the words stops short at the level of research policy for the individual provinces of knowledge. Beyond that, when the problem requires us to consider scientific knowledge as a whole and compare knowledge from one province with that from another, different considerations must come into play. The immediate question is, therefore, if the different provinces of knowledge are scientifically incommensurable, on what grounds can they be compared?

III

Any discussion at the national level on policy for scientific development is driven, sooner or later, to consider the question of purpose. At this level the problem of policy is not one of scientific merit or demerit, feasibility or unfeasibility, but of choice between broad possibilities all of which are in some way desirable but not all of which can be undertaken with the resources available. It is at this stage that the question of national purpose becomes the ultimate deciding factor for, in the last analysis, it is this that will determine the support that a country is prepared to make available for scientific development. We must be clear, therefore, as to what is meant by 'national purpose' and the factors that go to make this up.

As I see it, 'national purpose' rests primarily upon the existence of a consensus of opinion in a particular country on the desirability of attaining a particular objective. This differs from wishful thinking by having taken feasibility into account, for only in so far as there is confidence that an undertaking is feasible can it graduate from a wish to a purpose.

If this definition is valid—as I believe it is—then it leads to important conclusions regarding the support of scientific research in one province relative to another. Clearly, if society has no desire to achieve a particular objective then the feasibility of its attainment will be irrelevant. Equally, if some objective is strongly desired then its feasibility will become of decisive importance. On the other hand, if two different objectives are both scientifically feasible, the support that society will wish to be made available to each will be determined by the relative strengths of the national desire in regard to either.

The ultimate determinants of human conduct in regard to material things are not themselves sophisticated. They are what they have always been: the pursuit of an improved standard of living, security and health; and the stability of human communities primarily depends upon the ability of their organizations to satisfy expectations in these respects. Although, therefore, civilizations in the course of their elaboration have developed highly complicated techniques, such as economic systems, for achieving these objectives, to attempt to account for national purpose purely in terms of these would be to fall into the classical error of mistaking means for ends. These considerations are as relevant to the problem of national policy for scientific development as they are to any other problem of government.

There are many fields of scientific research which do not contribute to the production of marketable goods and which, in consequence, are not susceptible to meaningful expression in financial terms. As the common phrase 'a man of means' recognizes, societies do not seek wealth for its own sake but in order to have the means to obtain things they want. Thus, although as a practical necessity means must be acquired, the acquiring of these is essentially a secondary matter, and any attempt to base a social justification for scientific development solely on these considerations would be essentially to mistake the issue. Take biomedical research, for example. On purely economic grounds it might seem obvious that an underdeveloped country should devote the weight of its resources for research to agriculture and reduce to vanishing point its support for sophisticated research in medicine. Or alternatively, in an industrial country, dependent on its exports, logic might suggest that the brake should be applied to medical progress with its escalating demands on the national economy. As a matter of practical politics, however, the overriding consideration is that the motive force in national policy for medical research is not economic. It is the universal desire of men that they and their kin shall not die before they need and their determination not to be deprived of any research that promises to help them in this regard. It is essentially for this reason, not economic rationalizations, that nations are prepared to devote considerable resources to developing biomedical knowledge and even to forego, to some extent, other feasible objectives that on grounds of cold logic might seem to deserve prior consideration.

And standards change with time and place. All men seek to avoid hunger, but when hunger is satisfied they wish to feed better. Men wish to communicate and always faster. They want materials and increasingly more variety in these. What was luxury yesterday can become necessity today when scientific research by its achievements has created new standards of expectation. Over and above the basic level of plain want, such standards vary from country to country. If a society is particularly health-conscious, then, relative to others, it will lavish money on biomedical research; if accustomed to a particular level of security, on research related to military ends. Again, a country that depends for its external earnings on agriculture will be disposed to give a high priority to research that contributes to this, whilst one dependent on its manufactured exports will equally be disposed to favour research relevant to its industries. Similarly, a country faced with diminishing resources either in materials or energy will be inclined to seize on research aimed at producing alternatives; one with an excess of a particular resource to that aimed at exploiting its uses. And in times of national crisis, such as actual war, the whole system of desires and expectations may change radically.

These are all factors that are rooted in the primary needs of human communities. According to circumstances and changing conditions their relative strengths will vary between different communities and in the same community from time to time. But over and above these there are other factors that, when conditions are propitious, may enter substantially into considerations of purpose. Today, when the importance of scientific knowledge for national development has been conceded, scientific leadership has become a recognized element in national prestige and the ranking of countries in the hierarchy of nations. Quite apart therefore from cold objective assessments of utility, particular lines of research may acquire a significance in public regard that, from the point of view of national purpose, is nonetheless important because it rests on political rather than scientific or intellectual grounds.

The considerations that enter into the formation of such broad national wishes are, therefore, many. The assessment to be given to each in order to arrive at the strength of the resultant desire is a problem demanding its own expertise. Even more demanding is the comparison of the strength of the desire for one objective relative to another. But such comparison can be made. Broad objectives can be ranked in an order of priority on the basis of the relative strengths of the national needs in regard to each. It is against such a background that a national policy for scientific development has now to be seen.

IV

On first analysis, therefore, the problem at the level of national policy is that of reconciling the requirements of scientific developments on the one hand with those of a broad order of priority in national objectives at the other. The question is, can this be done? In this connexion it would be idle to close one's eyes to the fact that among a substantial proportion of scientists there is a deep feeling that this is not only impossible but that it is bordering on a betrayal of research for any scientist to entertain the possibility. To the extent that attempts to impose administrative or political direction on scientific research have in general been self-defeating and even on occasion damaging to the interests of all concerned, these fears are not entirely without justification. But it does not follow that because mistaken proposals have been, and continue to be, made in this connexion that the problem is necessarily insoluble nor that in addressing ourselves to it we are attempting to reconcile the irreconcilable. In my opinion, the problem is soluble and the solution is to be found along some such lines as the following.

Clearly it is only possible to reconcile two different sets of considerations if there are areas in which these come in contact with each other. In respect of an order of national objectives on the one hand and scientific development on the other such areas do I believe exist. They are to be found at the mission or purpose orientated frontier of each of the several provinces of natural knowledge. It is here, and only here, that national purposes and scientific development come into direct contact and through such points of attachment, and only through such points, that the individual scientific subjects can establish their relationship in the national setting. Thus molecular biology in the context of biomedical and bioagricultural knowledge, although at several removes from direct contact with national purpose, can be seen to be an essential part of the intellectual continua in question. Similarly chemistry can be seen to be part of the continuum related to materials, nuclear physics of that related to energy. Apart from the relevant context, none of these subjects can establish a more compelling claim to national consideration than that of its intrinsic interest. Within such contexts, however, they find their natural place and justification. In relating scientific development to national purposes, therefore, the units with which we have to deal are, not individual subjects, but whole provinces of natural knowledge.

These considerations point the way to the first and essential distinction that must be drawn. This is a distinction between research policy concerned with the development of a particular province of knowledge and national policy for scientific development which is concerned with the promotion of one province of scientific knowledge rather than another. The considerations entering into each of these now require attention.

Naturally, the central research organizations responsible for developing the different provinces of scientific knowledge must be clearly aware of the national purpose relevant to their particular concerns. This they must necessarily take into account when determining the relative emphasis that, within the limits of scientific feasibility, they will give to the development of the different sequences that collectively make up their particular province. A central research organization would, however, be denying its *raison d'être* and betraying its professional duty to society, if it ever allowed political or administrative considerations to override its scientific judgment of the way that knowledge in its particular province could best be developed. That is the reason why such organizations must be autonomous and under scientific direction and why they must, as the Trend Committee endorsed, be separated from the administrative functions of Government.[1]

At the level of research policy proper, therefore, the issues are

[1] *Committee of Enquiry into the Organization of Civil Science*, 1963, para. 61.

clear and the way to handle them evident. It is beyond that level the difficulties arise.

If one were asked to define the particular and essential problem that distinguished national policy for scientific development from that at all other levels, I think that it could be put this way. At the national level the distinguishing problem is that of determining, in the light of the resources available, the best balance to be struck between the competing claims of the different provinces of scientific knowledge so that decisions can be reached on the support to be given to each. To this end we need to take into account four different categories of consideration, to work out their interactions and to arrive at a final synthesis that comprehends them all. These four factors are respectively the political, the economic, the scientific and the administrative. It falls now to consider each of these in turn.

In considering the concept of national purpose and the feasibility of placing such purposes in a broad order of priority we have already identified the political element in the problem.

Economic considerations are of two kinds. The first relates to the rapidly rising cost of scientific research and its consequent claims on the national exchequer; the second to the economic benefits that may be expected from the development of projects within one particular province of knowledge in comparison with those in another. Thus, scientifically speaking, it may be perfectly feasible to build a costly machine for research on nuclear physics and equally possible to promote research towards a series of biomedical objectives. But it may not be possible to find finance to do both. A costly investment in one particular province of knowledge may promise great economic profit. Equally a comparable investment in biomedicine or bioagriculture may promise great savings. Again available finance may not permit both to be undertaken. In both instances economic considerations force a choice to be made and in so doing sharpen considerations in respect of priorities in national purpose and feasibilities in scientific research.

Scientific considerations at this level are again of two types. The first is relatively straightforward. It is concerned with the assessment of costs. The sums required for scientific research are now so large that it is only reasonable that governments should feel that they must have an independent assessment of the estimates put forward by those identified with particular projects so that they may feel assured that, scientifically speaking, these are in fact properly costed. The second type of problem is the difficult one and it is this that distinguishes scientific consideration at this level. It is that of comparing the relative promise of research in the different provinces of knowledge to attain their intended objectives. In this respect there are all degrees of feasibility between the virtual certainty and the outright gamble. Such degrees of feasibility can be broadly assessed in regard to each province of knowledge and, on this basis, the relative promise of one province to attain its scientific objectives compared with that of another. An artist may not be able to paint well in all styles but he is likely to have a better idea than a layman whether or not a picture in any style has been well painted.

Lastly there are the administrative considerations. Now, when the requirements for scientific development have such wide repercussions, the full range of their implications within the machinery of any social structure cannot be neglected. Informed opinion on these aspects is indispensable and it is in this respect that the expertise of the administrator is required in its own right.

At the national level, therefore, policy in respect of scientific development is the outcome of several categories of consideration. None of these is sufficient in itself but any one may override the others. The organizational problem at this level is, therefore, how to provide for the necessary synthesis of them all and so to obtain a basis on which national policy can be built.

V

It is evident today that scientific knowledge is relevant to a greater or less extent to the activities of all the departments of government. Further, that in respect of any one department this relevance is not confined to any one province of knowledge. It is equally evident that, as the various provinces of scientific knowledge develop, they are becoming increasingly interlocked so that any policy which segregated them from each other would be a misjudgment of the developing scientific situation. But this is by no means the whole problem. With increasing frequency scientific research is uncovering information that in its implications goes well beyond the straightforward technical considerations of specialized departmental policy. In the previous chapter we have given instances in which findings within particular provinces of knowledge were of major importance to the policy of Government itself. Further, in the preceding section we have seen that considerations at the national level must take in the whole span of scientific development. All these considerations point clearly to one conclusion. Although the inflow of information at the departmental level can be adequately catered for by the ordinary working relationships between these and the different central research organizations, the provision for scientific development and the inflow of information with major implications must come to one single focus; and this must be sited at the centre of government itself.

Doubtless it is the appreciation of the logic of the situation that lies behind the new arrangement in Canada by which the machinery for scientific development comes directly under the Prime Minister, and the arrangement in the United States of America in which the Scientific Advisory Committee is directly responsible to the President. Necessarily the exact arrangement in any country will depend on the social structure that it has evolved. The natural requirements to be met are, however, the same everywhere. In the context of this country I, personally, am doubtful about the wisdom of relating scientific development directly to the Head of State because of his inevitable preoccupation with a multitude of other cares. But I am entirely convinced that Haldane was correct in relating this to a senior Minister of Cabinet rank without departmental responsibilities. My suggestion in terms of the social organization in this country would be, therefore, as follows.

There should be a Ministerial Committee of Cabinet made up

of Ministers whose Departments are the most substantially concerned with scientifically-dependent activities. This should meet under the chairmanship of the senior Minister referred to above as designated to take over-all responsibility for scientific development and would constitute the focus from which authority would come and to which scientific information in its broad and important aspects would flow. The various central research organizations would be directly responsible to this Minister for Scientific Development in regard to the development and implications of knowledge within their particular province of knowledge. But this, although a major and necessary provision, is only part of the problem that will face the Minister and the Ministerial Committee for Scientific Development. There remains the essential duty of finally deciding the national policy for the relative development of the different provinces of scientific knowledge.

VI

The synthesis of the political, economic, scientific and administrative considerations required to arrive at a decision on national policy for scientific development is an exacting task. By the time that problems have reached this level it is only rarely that any one of such considerations will, in itself, point clearly to a conclusion. More often the upshot will be a series of possible courses of action each of which has its advantages and disadvantages. If, therefore, Government is to be in a position to make informed choices, it is necessary, now that considerations have become so technically complex, that suitable arrangements be made for casting up the various possibilities open for decision. To this end a suitably constituted advisory body for scientific development is required.

In principle, this need has been generally recognized, and in different countries various arrangements have been implemented to this end. It would probably be not unfair, however, to say that no country seems to be entirely satisfied that it has yet solved the problem. In general the arrangements that have been made seem to be open to one or other, or both, of two serious criticisms. The first is that in formulating the terms of reference of such bodies a sufficiently clear distinction has not been drawn between research policy (which we have seen is only feasible in respect of individual provinces of knowledge) and policy for the sciences which is the essential point at issue. As a result such bodies have tended to become embroiled in matters at the individual subject level and so to be distracted from their proper purpose. The second criticism is the consequence of not having ensured that all the necessary considerations—political, economic, scientific and administrative—are brought together in one body but rather allowed these to remain separate, in part or in whole, to the detriment of effecting that synthesis which is the major problem at this level.[1] Not all the arrangements proposed, however, have been of this kind. The recommendations of the Trend Committee for instance would, had they been accepted, have largely met the requirements. These provided that the necessary Advisory Council should consist, in addition to an independent Chairman, of some fourteen other members, half of whom should be scientists and half made up of industrialists, economists and individuals with wide experience of public affairs.[2]

Clearly if any deliberative body is to fulfil its purpose then its membership must, as the Trend Committee recognized, reflect the range and experience of the matters which come before it. On this basis, therefore, an appropriate composition for an advisory body concerned with national policy for scientific development might well be something like the following.

The suggested composition of an independent chairman and fourteen members, half of whom would be scientists, seems about right. The non-scientific members should include, as proposed, some eminent industrialists, some with wide experience of public

[1] 'Any organization of science must be a comprehensive task and one which cannot be undertaken alone either by scientific workers themselves or by the State or economic organizations outside science, but only by all working together in an agreed direction.' J. D. Bernal: *The Social Functions of Science*, Chapter XI, p. 241. George Routledge & Sons, Ltd, London, 1939.

[2] *Loc. cit.*, para. 113.

NATIONAL POLICY AND SCIENTIFIC DEVELOPMENT

affairs and an economist. To these, however, I should propose adding a man with administrative experience of the machinery of government. The scientific members would fall into two categories. The first would consist of three or four independent scientists one of whom would be an officer of the national academy of sciences or its equivalent. The second would consist of the chairman of the previously suggested Chief Executives Committee of the several central research organizations and two other members of that committee so that the three together covered a representative selection of the provinces of natural knowledge. This latter suggestion is an extension of the Trend proposals.[1] It is evident, however, that as the Advisory Council will need to be concerned with the plans and premise of the operations of the central research organizations they will need to have first-hand information on the ways that these function. Finally, a representative of the Minister for Scientific Development would naturally attend meetings of the Council so that relevant information on Government policy could be available.

VII

Up to this stage, discussion has been confined to those features of national policy for the sciences that distinguish this from research policy as such and provide the reasons for organizing specially to this end. There remain, however, three aspects of scientific

---

[1] Again I have to confess to an apparent change of view. The personal opinion I expressed to the Trend Committee was that the chief scientific officers of the central research organizations (like the Chairman of the University Grants Committee) should not be members of any advisory council to the Minister at the level of general policy for scientific development. My reason was that they would thereby compromise the independence of their own organizations which should continue to be free to advise the Minister directly on the matters in which they were more informed than any general body could ever be. Such would still be my view if any general advisory body were at liberty by its terms of reference to concern itself with research policy proper within the field of responsibility of particular central research organizations. If, however, the distinction between research policy and policy for the sciences were drawn and a body orientated to the problems of this latter set up, my view would be otherwise.

activity that still require consideration. The first deals with defence research, the second research involving international co-operation, and the third concerns scientific manpower.

Defence research, from its nature, must often be confidential and its results not generally available to the scientific community. Yet much of the research related to defence has implications in the civil sphere and scientific research for civil purposes has implications for defence. Clearly, unless substantial reduplication is to occur and interchange of relevant skills and knowledge to be inhibited, some means of bridging the gap is needed. Normally, the chief scientific executive officers of central research organizations are aware of activities in the defence field that are relevant to research in the provinces of knowledge for which their particular organizations are responsible. On this basis, working relationships can be expanded or reduced according to national circumstances. But there is also a need for general awareness of the scientific situation in the defence field on the part of any advisory body concerned with national policy for the sciences. Doubtless the responsible Minister would ensure that the necessary background information on defence research is suitably made available.

International scientific research may be undertaken for purely scientific reasons or other considerations of a primarily political nature may come into account. Scientifically speaking, the case for international co-operation is incontrovertible when dealing with fields of experience that can only be studied on a world scale or when the costs of particular research are so great that they can only be met by countries combining their resources. Examples of the former are the study of epidemiological problems such as the spread of epidemics of infectious diseases, or of conditions that occur too infrequently in any one country for it to be possible to study them in any unit smaller than the world. Examples of the latter are research in nuclear physics and in the exploration of space. Once agreed, research projects of these kinds will normally fall to be dealt with operationally by the central research organization responsible for the province in which the subject in question falls. But international co-operation always

NATIONAL POLICY AND SCIENTIFIC DEVELOPMENT

raises questions of national politics. Indeed, on occasion, the main case for a particular international project may be that it is important for political or ideological reasons. Among the exacting problems that may be referred to any advisory body on the balance of civil research are those in the international sphere. Provided that there is a genuine scientific reason for the particular project being administered internationally, such questions give rise to little dubiety. If there is not, its support may well be mistaken policy on any count. Further, there is the general problem of international exchange as distinct from collaboration. National academies have a major role to play in this respect, but, as governments are frequently involved either directly or indirectly, academies may need to co-operate with the governmental organizations for scientific development.

Research and development depending as they primarily do on the endeavours of trained men, any country concerned with its progress in scientifically dependent matters must inevitably pay close attention to the problem of scientific manpower. But only to a limited extent can this problem be considered by scientists alone. Its full solution also involves considerations of a country's educational policy. Whilst, therefore, it falls to the responsibility of scientists to determine the needs for trained men in the different aspects of scientific work, it must equally fall to the educationalists to determine the implications in terms of educational policy of meeting these. Considerations are required at all levels. All central research organizations are necessarily concerned with the problem as it affects the development of their particular province of knowledge. Collectively, these assessments raise the problem to the level of national policy, and it is at this level that scientific considerations need to be married with educational. Clearly this is a matter that in the last analysis only Government can decide. There can be little doubt, therefore, that any advisory council on scientific development will need to concern itself with the national policy for scientific manpower.

VIII

In the preceding considerations we have approached the problem of national policy and scientific development largely from the standpoint of the scientific community. It would, therefore, be as well if, before leaving the matter, we tried to see it from the opposite point of view. In other words, we might ask ourselves, what does a modern society, as expressed by its government, now need from its scientists over and above their own personal contributions to research? I think that essentially it needs three things. It needs effective organizations to promote research in the various provinces of scientific development. It needs effective arrangements by which the implications of developing scientific knowledge are authoritatively brought to the notice of its government and by which problems that depend on scientific knowledge for their solution can be given informed assessment. It needs advice on the deployment of its resources between the different provinces of scientific endeavour. We can now summarize our considerations in each of these respects.

The units of organization for the promotion of research are the various central research organizations each corresponding to a province of natural knowledge which stretches over the whole span of experience from the most specialized to the increasingly unspecialized. It is on these that society must rely not only for the development of knowledge in the particular province of natural experience but also for the assessment of the extent to which such knowledge bears on the achievement of the various scientifically-dependent purposes that the society in question is concerned to realize. These are professional tasks and, in consequence, such organizations must be executive and autonomous and controlled essentially by scientists. Integration with the machinery of government is necessary at two levels; at the departmental and at the centre of government policy itself. At the departmental level integration is secured by the establishment of working relationships with the several departments of government to the specialized interests of which knowledge in the particular province is relevant. At the level of government itself relationships are concerned not with interests confined within the remit of particular departments but with the implications of developing knowledge in the particular province for governmental policy that involves several departments or is of general national concern. For this reason (and also because of the diversity of departments to which knowledge in any particular province is, or may be, of relevance), it is essential that central research organizations be related directly to the centre of government itself, and not segregated in any particular government department as conventional thinking might, at first sight, suggest.

To this end there should be a Ministerial Committee of Cabinet for Scientific Development under the chairmanship of a senior Minister without specialized departmental responsibilities; and to this Minister the several central research organizations should be directly responsible.

It is, therefore, to its confederation of central research organizations that a country must look not only for the actual promotion of scientific research but also for expert appraisal in regard to the implications and possibilities in the various provinces of scientific knowledge. Without the informed opinions and machinery for executive action that organizations of this kind provide, government policy for scientific development would, in modern circumstances, be impossible to realize.

But the essence of government is choice and, now that scientific knowledge has become of such importance to national purpose, the proper deployment of a country's resources for scientific development between the several provinces of natural knowledge has become a problem of major importance. To regard this as something that can be solved on the basis of scientific considerations is to disregard both the nature of the problem and the nature of scientific knowledge itself. But if national governments are to make informed decisions on this problem they must have informed advice in so far as its technical aspects are concerned. In regard to the political element in these considerations, that is a matter for government itself. In regard to the permutations and combinations of the economic, administrative and scientific elements, that is a matter for the combined skills of experts in these lines. It is in connexion with this latter and special requirement that the need arises for an advisory body to the Ministerial Committee for Scientific Development. In principle, this is now widely recognized. But if such a body is to fulfil its purpose its remit must be clearly focussed and its composition reflect the range of considerations that enter into the solution of the problems with which it has to deal.

* 13 *
EPITOME

THE INCENTIVE TO WRITING THIS ESSAY was the belief that scientific development is now entering upon a further era in its evolution and that, if the requirements of this were to be adequately met, new thought would need to be given to its organization. Essentially the factor that has brought this situation into being is the growth of scientific knowledge itself and the resulting realization by society of its increasing dependence upon this. But the situation is not unprecedented. It is a recurring phenomenon of social evolution that as knowledge grows it compels society to devise further measures for its deployment and promotion. In the course of history, under the pressure of such growth, we have seen the successive emergence of specialized professions and occupations, the universities, the scientific academies and, more recently, the nation-wide central research organizations. Now we are faced with the problem of national, and international, organization for scientific development and the adjustments to the devices already in existence required to bring all into an integrated whole. But natural phenomena are the same always and the conditions required for their successful investigation, or exploitation, remain unchanged. If, therefore, our administration is to be more than a succession of temporary expedients, we must take care that it conforms to the realities to which it seeks to give expression. It is for this reason that I have felt that the first requisite in facing the situation now opening before us is to re-examine our ideas on the structure of scientific knowledge in the light of the developments that have occurred since our traditional views were formed. Given a correct understanding of this, the administrative measures appropriate to the new circumstances should be self-evident and effective. Without it they are condemned to expediency. I am aware that, in the examination that I have set out in this essay, I have questioned many cherished traditions. I am also aware that, for this reason, I have been driven to treat some aspects of the problem disproportionately. It is useful, therefore, at this final stage to attempt to see the problem as a whole and in perspective.

Despite the different views that have been put forward on the structure of scientific knowledge, there is general agreement that it orginated from enquiries into natural phenomena started in many different places. Only by drawing an arbitrary distinction between the knowledge gained in such primitive specialized enquiries and that gained in their further development[1] is it possible, however, to conceive of this latter giving rise to a single intellectual entity called 'science' which on further development divides up into the various branches of scientific knowledge. The reasons why the attitudes of men's minds were predisposed to adopt such a concept have been suggested, and rejected. Instead, agreement has been expressed with Karl Pearson's dictum that the only unity in scientific knowledge is the methodology by which it is gained and that the knowledge itself is as various as the natural experiences from which it is derived. On this basis I have been emboldened to suggest a different concept of structure: that of a vast globe of ignorance from the surface of which at many different places enquiries are being driven centrally to deeper levels where they tend increasingly to coalesce and to produce subjects that come to underlie, not one, but several such penetrations. These penetrations I have called provinces of natural knowledge, and I conceive of each of these as forming a smooth continuum of interrelated subjects from its specialized periphery to the more unspecialized depths of its penetration. It is within the perspective of such provinces that the individual scientific subjects find their intellectual context and larger significance. It is because of the intellectual relationship between subjects within these that it is possible to devise, on the basis of genuine scientific

---

[1] 'For it is drawn either from the mechanical arts, or from a number of crafts and experiments *which have not yet grown into an art properly so called*. . . .' Bacon: *The New Organon*, p. 277.

considerations, a comprehensive research policy for their further development.

But between the knowledges characteristic of each of the different provinces it is impossible to demonstrate a similar scientific relationship. It is true, of course, that as the different provinces of knowledge develop towards the more unspecialized central regions they come to have elements in common. But considered in general the different provinces of natural knowledge are, scientifically speaking, incommensurable. The idea that it is possible to devise a policy which would embrace the development of scientific knowledge in all its aspects is, therefore, devoid of scientific foundation.

The administrative implications of this concept and these considerations are evident. For each province of natural knowledge an organization is required to develop comprehensively the full range of intellectual interrelationships which this represents. Within this range research policy in a genuine scientific sense is feasible and, to this end, central research organizations operating on a nation-wide basis have evolved. Beyond this level, however, considerations change. Now the problem is to evolve a basis upon which one province can be compared with another so that support can be adjusted to accord with the different national aims. Now the relevant considerations are political, economic, administrative, and scientific only in so far as these bear on the assessment of the promise of a particular province of knowledge to attain its intended objectives. Ultimately decisions at this level are governmental. So pervasive is scientific knowledge at this present time, however, that these must come to a focus at the centre of government itself. Accordingly it is suggested that a ministerial committee of central government, under the chairmanship of a senior Minister, is required. But if informed decisions are to be taken on this complex problem, government must be advised. To this end an advisory body, the membership of which appropriately represents the range of considerations involved, should be created.

As I see it, therefore, in devising our organization for scientific development we must clearly distinguish between research policy and policy in a particular country for the support of scientific activities. In regard to research policy, scientific considerations take precedence over all others. Only in so far as a country has the wisdom to ensure that it has continued access to scientific assessments unbiased either by wishes or the commitments of previous policy can it, in the present technologically-dependent age, feel justified in facing the future with confidence.[1] To this end, as Haldane saw long ago, the several central research organizations responsible for developing knowledge in their different provinces must be autonomous and independent of the administrative machinery of government. In regard to policy for the support of scientific activities in a particular country, the ultimate decisions are necessarily matters for government. As such they must be based upon a synthesis of considerations; political, economic, administrative and scientific in so far as these latter concern the relative feasibilities of realizing particular national purposes. To facilitate this, an advisory body to government, with a composition that suitably reflects these various considerations, is now indispensable. But this must be clearly distinguished from the essentially scientific bodies responsible for promoting the development of knowledge in the different provinces and for making this available.

It has been a central theme in this essay that for any system of organization to be effective it must continue to satisfy two basic requirements. The first is that it must be in conformity with the deeply held sentiments of those who have to make it work, for only thus can it be productive of the necessary loyalty and initiative. The second is that it must be equally in conformity with the realities of the situation that it seeks to organize. Scientific research is directed to the understanding of the unknown. It is natural phenomena themselves that pose the essential questions and determine the limits within which we can

[1] Although informed opinions may differ on the interpretation of the actual details of the incident around which C. P. Snow constructed his Godkin Lectures (the pre-war dispute on the value of radar as an early warning system), none could question the significance of the situation he considers for a modern government. C. P. Snow: *Science and Government*. Oxford University Press, 1961.

operate. Only to the extent that our organization continues to express our appreciation of this will it be effective and reliable. It has been the purpose of this essay to attempt such an appreciation and to translate this into terms of organization within the context of the new era into which scientific development is now entering.

INDEX

Addison, Christopher
  on medical research, 102-4, 150
Administration
  scientists and, 138, 143-4
Administrative Departments
  relation to Research Councils, 101-103, 148-50
Aerodynamics, 42
Agriculture, Province of, 27-9
Analogies
  see models
Applied Research, 53-6
Atomic Bomb, 1
Autonomy, University
  national significance, 79
  research and, 93-6
  safeguard for, 95-6

Bacon, Francis
  scientific knowledge
    genesis of, 48, 51
    structure of, 48
Basic Research, 18-19, 53-6
'Big Science', 56-7
Biochemistry
  genesis of, 15-16
Biomedical research
  progress in, 7-8
  sequences of, 9-12

Cancer Research, 10
Categorization of Knowledge
  proposed, 59-61
  traditional, 53-6
Central Research Organizations
  administration of, 142-5
  autonomy, 99, 104, 174
  collaboration between, 145-8
  control by scientists, 127, 129, 142
  deployment of resources, 130-41
  dual role, 142
  genesis of, 72-3, 98-9
  kinds, 109-14, 140-1
  own staff, 132-4
  principles of, 99-104
  relations to
    administrative Dept., 148-51
    central Government, 151-2
    scientific basis, 104-5, 113-14
    social responsibilities, 136-8
Chemistry
  genesis of, 18-19
Chemistry (Materials)
  province of, 30-5
Chief Scientific Executives, 143-5
  committee of, 148
Clinical Research
  centralized, 117
  decentralized, 117
  operational, 117-18
Contexts of Knowledge, 25

Defence Research, 166
Department of Scientific and Industrial Research (DSIR)
  alternative possibility, 109-12
  dissolution of, 108
  replacement of, 108, 120
Developmental Research, 53-6
Directing Research, 24, 125

Earth Sciences, 29-30

INDEX

Education of Research Workers, 89–93
Energy (Physics)
  province of, 35–42
Evolution of Organized Research
  academic, 68–9
  background to, 65–7
  national organizations, 72–4
  practical, 69–71
  societies, 71

Freedom, academic, 69, 78–9
  justification for, 78, 93
  research and, 93–6
  safeguard of, 79, 95–6
Fundamental research, 18–19, 53–6

Graduate Universities, 89
Gresham College, 70

Haldane of Cloan
  'Haldane principles', 100–1, 150, 174
  machinery of Government, 100–1

Industrial Research
  centralization, 120–1
  decentralization, 119
  massive projects, 122–3
Instruments for Research
  academies, 135
  industry, 119–23, 131–2
  national research organizations, 130–4
  professions, 117–18, 131–2
  societies, 134
  universities, 73–97, 130–1
International Research, 166–7

Jet Engine, 42
Justification for Research Support, 57–9

Knowledge, Scientific
  categorization
    proposed, 59, 61
    traditional, 15, 44
  contexts of, 25
  for own sake, 23–4

specialized, 13
unspecialized, 13

Levels of
  motivation, 23–5
  research, 21–2
'Little Science', 56–7

Materials (Chemistry)
  province of, 30–5
Mathematics, 18
Medical Research Council (U.K.)
  origins of, 99–200
  principles underlying, 101–4
Minister for Scientific Development, 150, 163
Ministerial Committee for Scientific Development, 152, 162–3
  advisory body to, 163–5
Mission-orientated Research, 19–20
Models, conceptual
  globe of ignorance, 50, 172
  implications, 51–9
  tree of knowledge, 47–9
Molecular Biology
  genesis of, 17
Morale, 103, 104, 174
Morant, Robert, 1, 99, 138
Motivation in Research, 23–5

National Environment Research
  organization for, 112
National Purpose
  concept of, 155–8
  factors involved
    administrative, 161
    economic, 160
    feasibilities, 161
    political, 160
  research policy and, 158–60

Operational Research, 100, 117–18
Organization
  alignment with reality, 171–2
  basic requirements, 104
  morale, 104

Pearson, Karl
  on unity of science, 52

INDEX

Ph.D., restrictions on, 92–3
Physics (Energy)
　province of, 35–42
Policy
　national, 158–63, 168–9, 173
　research, 124–9, 154, 172–3
　'scientific', 154
Politics and Research, 155–8
Postgraduate Education and Research, 88–9
Practice and Research
　organization of, 115–16
Prestige Research, 157
Private Research Foundations, 135
Provinces of Knowledge, 47, 50
　and central research organizations, 113–14
　kinds, 109–13, 140
　units of organization, 140–1
Pure Research, 18–19, 53–6

Radar, 40
Radioactivity, 41–2
Research Councils
　see Central Research Organizations
Research Policy
　basis of, 128–9, 140–1
　control by scientists, 129, 138–141
　directing workers, 125
　integration with Government, 168–170
　limitation of, 173–4
　and national purpose, 158–60
　possibility of, 124–5, 154
Research Training Schemes, 167
Research in Universities, 73–97
　academic freedom and, 78–9
　contribution from, 76–81
　dual support of, 81–4
　entitlement to support, 81
　initiative in, 78–9
　limitations of, 97
　policy for, 93–6
　sophistication, increasing, 84–6
Rockets, 42
Royal Institution, 70

Royal Society, 70, 135

Science as a Unity, 51–3
Science Research Council (U.K.), 108
Scientific Policy, general
　feasibility of, 154, 173
Self-orientated Research, 19–20
Sequences of Knowledge
　basis for research policy, 128–9
　characteristics, 13–15, 46–7
　component qualities, 44
　historical growth, 43
Sociology
　organization for research, 113
Specialization, 20–5
　bias from, 21, 24–5
Spencer, Herbert
　genesis of scientific knowledge, 44
　structure of scientific knowledge, 49
Subjects of Knowledge
　context of, 20
　mission-orientated, 19–20
　self-orientated, 19–20
Support of Research
　academic, 81–4
　escalation of, 1
　justification of, 57–9

Thermodynamics
　genesis of, 36–7
Transportation
　organization of research, 41–2
Tree of Knowledge
　derivation, 47–9
　implications, 51–6
Trend Committee, recommendations on
　advisory body on development, 163–5
　DSIR, 105–7, 119–21
　research councils, 103–4
　scientific executives, 144–5

Undergraduate Teaching and Research, 86–9
Understanding for its own sake, 23–4

INDEX

Unity of Science, 51–3
Universities
    Graduate, 89
    Undergraduate, 89
University Grants Committee
    (U.K.), 81–4, 94–5

University of London, 72
Universities and Research
    see Research in Universities

Weinberg, Alvin
    on specialization, 21